The Scars of Evolution

THE SCARS OF EVOLUTION

Elaine Morgan

'Remnants of the past that don't make
sense in present terms – the useless, the
odd, the peculiar, the incongruous – are
the signs of history.'

Stephen Jay Gould

OXFORD UNIVERSITY PRESS
NEW YORK   OXFORD

Oxford University Press

Oxford   New York   Toronto
Delhi   Bombay   Calcutta   Madras   Karachi
Kuala Lumpur   Singapore   Hong Kong   Tokyo
Nairobi   Dar es Salaam   Cape Town
Melbourne   Auckland   Madrid

and associated companies in
Berlin   Ibadan

Library of Congress Cataloging-in-Publication Data
Morgan, Elaine.
The scars of evolution / Elaine Morgan.
p. cm.
Originally published: Souvenir Press, 1990.
Includes bibliographical references and index.
ISBN 0-19-509431-X (Pbk.)
1. Human evolution. 2. Body, Human. 3. Physiology, Pathological.
I. Title.
GN281.4.M68   1994
573.2—dc20   90-3525

2 4 6 8 10 9 7 5 3 1
Printed in the United States of America

Acknowledgements

This is the third book I have written promoting a view of human evolution hitherto regarded as heresy.

I should like to express my sincere gratitude to Dr Marc Verhaegen. It was a paper published by him in 1987 which gave me the idea of approaching the subject from a different angle, and I have benefited greatly from his advice and co-operation during the writing of this book. (This does not necessarily imply that he agrees with all the opinions expressed in it.)

I am especially grateful to the family of the late Leon p. La Lumiere, Jr, for allowing me to reproduce his maps of the geological history of the Danakil Alps in Ethiopia, showing how our ape ancestors could have been marooned on an island.

My thanks are also due to the hundreds of people who have written to me since 1972 expressing interest, offering ideas and suggestions or enclosing cuttings and references. Their letters have been a constant and invaluable source of encouragement.

Contents

tower — *Kinks in the spine* — *Stresses on the ver-*
tebrae — *Disc trouble* — *Remodelling the muscles* —
Inguinal hernia — *Gravity and the blood* — *Varicose*
veins — *Haemorrhoids* — *Hypertension.*

List of Maps

These maps were prepared by the late Leon P. LaLumiere, Jr, of the Naval
Research Laboratory, Washington, D.C., to illustrate his theory of the location
for the evolution of bipedalism. Figures 1–4 were previously published in The
Aquatic Ape by Elaine Morgan (1982). Figure 5 (after Hutchinson and
Engels, 1970, 1972) was previously published in GEO Magazine, 14
November, 1984.

Introduction

'What a piece of work is a man! How noble in reason!
how infinite in faculty! in form, in moving, how
express and admirable! in action how like an angel!
in apprehension how like a god! the beauty of the
world! the paragon of animals!'

Shakespeare: *Hamlet*

All general practitioners with heavy casebooks, reading Hamlet's eulogy, must be acutely aware that there is another side to the coin.

Since all their patients are mammals — even though of an exceptional kind — it is only to be expected that in common with the rest of the animal kingdom they may occasionally suffer from fractures and lesions, digestive disorders, birth defects, viral and bacterial infections and, in old age, impairment of functions.

But in addition to that, there will be some in every doctor's waiting room complaining of troubles which are unique to our species. They are apparently due to design defects — flaws in the human blueprint — the scars of our evolution.

It is as if at the birth of humanity (as at the birth of the princess in the fairy tale) good fairies clustered around the cradle bearing gifts. 'I endow this child with a wonderful brain and the powers of reasoning and inspiration . . .' 'I bring dexterity unparalleled in any other living thing . . .' 'I bestow the power of speech . . .'

But last of all came the bad fairy, and added the spiteful postscript:

'I curse this child with the propensity to suffer from lower back pains, obesity, enlarged adenoids, acne, varicose veins, cot deaths, sunburn, sleep apnoea, gynaecological and sexual malfunctions, dandruff, inguinal hernia, haemorrhoids.'

These and other things are the price we pay for being human. They came as part of the package deal.

So what sort of deal was it? What conceivable evolutionary path left us lumbered with these unwelcome gifts as well as the shining attributes praised by Hamlet? It is a fascinating riddle, and in searching for the answer it is necessary to begin at the beginning, with the most fundamental question of all.

1

The Emergence of Man

'The fact that an opinion has been widely held is
no evidence whatever that it is not utterly absurd.'
Bertrand Russell

Darwin's theory of evolution propounded an answer to
one major mystery about our species, namely, why we
bear such a powerful physiological resemblance to the
African apes — the gorilla and the chimpanzee. He argued
that it was because we and they share a common ancestor.

This solution immediately raised a further problem. If
we are so closely related to them — and everything we
have learned since suggests that the relationship is even
closer than Darwin supposed — then why are we not
more like them? When a species splits and gives rise to
three separate lineages, there is no reason to expect that
species C will differ from A and B any more widely than
A and B differ from one another. Yet in the case of the
anthropoids this is what happened. It is not because we
split off earlier: we didn't. Evidence from molecular
biology indicates that if any of the three diverged in
advance of the other two, it was the gorilla.

Darwin was well aware that he could not answer all the
questions surrounding human emergence. But his supporters were confident that he was on the right lines and
that it would only be a matter of time — perhaps a couple

1

of decades — before most of the remaining problems were solved.

One hundred and thirty-eight years have now elapsed since the publication of *The Origin of Species*, and scientists are no nearer to agreement on the question of why there are gorillas and chimpanzees and people, instead of merely three different species of African ape. Some people fail to see the point of the question. They would answer it by saying that we have simply 'evolved further' than the apes; that we have climbed a few more steps up the evolutionary ladder, while gorillas and chimpanzees have for some unknown reason lagged behind.

The concept of the evolutionary ladder is deeply rooted in our culture. It appealed especially to the Victorians with their belief in Progress and Perfectability, and it lasted a long time.

In the 1920s Elliot Smith, an influential British authority on human origins, lectured about 'Man's ceaseless struggle to achieve his destiny', and declared that 'for untold ages Nature was making her great experiments to achieve the transmutation of the base substance of some brutal ape into the divine form of man.' In the 1930s another pioneer, the Scots fossil-hunter Robert Broom, asserted: 'There was no need for further evolution after Man appeared.'

Even today there is widespread belief that we represent a kind of pinnacle to which the evolutionary process was aspiring; that 'more like people' equals 'higher up the ladder'; that *Homo sapiens* was in some way the object of the whole exercise, and that anywhere in the universe where the evolutionary process has gone on long enough, the emergence of an 'intelligent life form' akin to ourselves is a foregone conclusion, as a culminating triumph for the life force.

But the ladder is a myth. 'Nature' — the evolutionary process — does not aspire to anything. It does not aspire to intelligence in its creations; it does not aspire to complexity. These qualities in living things may emerge and be intensified — as it were under duress — when conditions of life become exacting. But simple and mindless organisms persist and multiply and thrive unchanged over many millions of years while more complex and powerful species like the dinosaurs rise and fall. Sentience, mobility and intelligence, like wings or eyesight or a sense of smell, can regress in a species where they no longer conduce to survival.

The essential point is that evolution takes place in response to things which *have happened*, not things which are predestined to happen. Man is no more an evolutionary pinnacle than a tree is, or a termite or an octopus. His emergence was no more inevitable than that of any other species.

We find it easier to accept that other life-forms are fortuitous, arising as a result of special sets of circumstances. For example, in some parts of the world geological changes have left freshwater fish marooned in underground caves. If that had not happened at some time in the past, there would be no blind white cave fish. In Australia there was once a primitive mammal that burrowed in the earth like a mole. If its territory had not at some time become increasingly waterlogged, there would be no duck-billed platypus. The world would have spun on and the biosphere would have continued its permutations without them.

The presence of people on earth is just as fortuitous. It might very easily never have come about. It takes a lot of explaining. Life on earth could well have proceeded on its way without us. In fact, nowadays Greens are tempted to say that the appearance of this improbable naked biped among the earth's fauna was not so much an evolutionary pinnacle as an ecological disaster.

Other people, while accepting that the presence of people on the planet requires an explanation, take it for granted that the explanation must have been discovered long ago.

This impression is understandable. Numerous books are published on the subject, written by professional scientists and aimed at the general reader. Several highly acclaimed documentary series have been televised dealing with evolution in general and/or human evolution in particular. There are A-Level courses where the topic is covered in some detail. Very rarely in these books and programmes and curricula is there any indication that a major and crucial problem remains unsolved. Usually the evolutionary narrative progresses smoothly, confidently and seamlessly from unicellular organisms to modern man.

There is usually a bland reference to the fact that the ancestral hominids began to stand up and walk about on their hind legs, as if it were the most natural thing in the world for a mammal to do, though no other mammal has ever done it. The question of why our bodies are hairless used to be discussed at some length, but now there is a growing tendency to avoid all mention of it. The writers have lost confidence in the answers they thought they knew thirty years ago, and naturally in addressing the layman they are not anxious to shine a spotlight on problems which have not been solved.

This motive does not operate in the case of the academic journals. Within those precincts the writers are not concerned with papering over the cracks. Almost all of them have an entirely genuine concern with discovering the truth. They therefore respect facts, identify sources, rigorously define the borderline between what is known and what is guessed at, and gracefully admit defeat when one of their beautiful hypotheses is slain by an ugly fact. In these journals an unsolved problem is not something to be brushed under the carpet: it is regarded as a challenge and an opportunity.

In the whole history of evolution, the emergence of

man is the one episode where the unanswered questions are thickest on the ground, and the challenge is most frequently taken up. At regular intervals, hypotheses are published attempting to account for one or other puzzling feature of human anatomy. These efforts may be criticised or they may be ignored. In either case the hoped-for chorus of 'Eureka!' fails to materialise and the theorist goes back to the drawing board. After a time another hypothesis is put forward, with the same result.

The present state of play may be summarised as follows. Four of the most outstanding mysteries about humans are: (1) why do they walk on two legs? (2) why have they lost their fur? (3) why have they developed such large brains? (4) why did they learn to speak?

The orthodox answers to these questions are: (1) 'We do not yet know'; (2) 'We do not yet know'; (3) 'We do not yet know', and (4) 'We do not yet know'. The list of questions could be considerably lengthened without affecting the monotony of the answers.

Scientists do not normally sum up the situation as starkly as that, for several reasons. In the first place, they do not see it like that. They have lived long enough with the conundrums to be no longer surprised by them; they are satisfied that new data are continually being amassed and steady progress is being made. That is quite true. As in any other form of enquiry, if you can produce solid reasons for discarding a false hypothesis, that represents a significant advance. Over the last thirty years a number of seductive evolutionary scenarios have been successfully demolished.

There is one other reason why they underestimate the extent to which they have been bogged down over this question for more than a century. That is the information explosion. Science has become ever more subdivided, and researchers have become absorbed in their own specialities.

Thus a palaeontologist on an African dig may be

obsessed by the problem of bipedalism and regard it as the last of the great unanswered questions. Crack this one, he feels, and we are home and dry. Meanwhile, in other parts of the world, other specialists are telling themselves that bipedalism must be pretty well understood by this time, and the really baffling characteristic still unexplained is the unusual nature of our skin — larynx — sex organs — sweat-glands — blood — adipose tissue — brain — tears — vocal communication — extended life span — or whatever is their own particular field of study.

In short, the chief mystery does not lie in any one of these anomalies, not even the wonderful brain or the dexterous hands or the miracle of speech. It lies in the sheer number and variety of the ways in which we differ from our closest relatives in the animal kingdom. Some of these differences are major and some are minor; several appear arbitrary and pointless; others are embarrassing or inconvenient; and it has not proved easy to trace any logical connection between them.

Something happened. Something must have happened to our own ancestors which did not happen to the ancestors of apes and gorillas. In the last analysis no one seriously disputes that. No one, for instance, expresses surprise that the anomalies were not more evenly shared, so that gorillas learned to speak, people walked on two legs, and chimpanzees became naked.

Something changed in the environment, or in the mode of response to the environment, or in the genes; that change affected only one particular group among the ancestral anthropoids, and set them off on the long road to becoming human.

The key question, then, is: 'What happened?' One possible response is: 'Who cares what happened? Whatever it was, it was millions of years ago and it can't possibly affect me.' But that is not true. Whatever it was, it affects every one of us because it had far-reaching consequences on the way our bodies function — or, frequently, malfunction. Possibly 20 per cent of the ills that our flesh is heir to can be traced back to the critical period of transition from non-human to human.

Theories about that turning point in our evolution cover a very wide spectrum, even if we omit the views of the Creationists who do not believe we evolved at all.

On the wilder shores of speculation is the kind of scenario which involves the incursion of beings from outer space bringing the benefits of their own acquired wisdom. This genre is an offshoot of the myth of the evolutionary ladder. It does not merely assume that some form of life has arisen elsewhere in the universe, which, considering the size of the universe, is a very strong probability. It also implies that as soon as molecules combine into amino-acids they are channelled along predestinate grooves which must end in creatures endowed with powers of independent mobility, reasoning, and manipulation, able and eager to build space-ships and cruise the galaxies.

That is infinitely less likely. It is also intellectually unsatisfying because it does not solve the problem of why human-like animals evolved; it merely relocates it in time and space. If we cannot work out how and why man evolved on our own familiar planet, it is not useful to refocus our enquiries onto how and why some alien analogue of ourselves emerged from an extraterrestrial precognitive life-form many light-years away.

At the opposite end of the spectrum is the orthodox school of thought which contends that nothing remarkable happened. One population of ape-like anthropoids migrated from a forest habitat to more open ground. And that was it; that was allegedly enough to set in train the series of changes that made us what we are today.

The strength of this idea lies in its near-inescapable logic, which runs as follows: (1) there are good grounds for believing that our very early ancestors lived in trees; (2) there are also grounds for believing that our less distant ancestors hunted big game on the plains of Africa; (3) therefore we know that our predecessors moved out from the trees to open country — that is, onto the African savannah.

This reasoning is almost foolproof, but not quite. If you see a man at the north end of London Bridge and then see him at the south end, you may feel safe in declaring that he crossed the bridge. However, if your second sighting is five years after the first and you notice that in the interim he has acquired a deep tan, a UCLA T-shirt and an American accent, it crosses your mind that he may have made some kind of detour.

The latter analogy is a closer parallel to what really happened. The earlier hominid fossils are of a creature utterly changed from an arboreal ape, walking upright on two legs. Previous to the appearance of these creatures there is a blank — a fossil gap which lasted for millions of years.

One serious weakness in the savannah theory lies in the disproportion between the commonplace event it depicts and the spectacular consequences alleged to have flowed from it. Many primate species such as baboons and vervet monkeys and patas monkeys have left the trees to live on the savannah. They have lived there quite as long as our ancestors ever did, yet they show no signs of acquiring any human-like features such as hairlessness or bipedalism. They settled down in the new habitat with scarcely any physical modifications, and there is no good reason why a cousin of a chimpanzee could not have done the same. Then there would have been two forest apes and one savannah ape, but no people to write books about any of them.

Despite these objections, the savannah theory continues to command the tentative allegiance of many of the best minds in the business. (It is only natural that the message getting through to most of the students is the allegiance rather than the tentativeness.) It is not that they are happy with it as it stands; there are far too many pieces that will not fit into the jigsaw. One subconscious factor may be a kind of curiosity fatigue, and a desire to work on different sectors of their field which yield answers rather more readily.

A recent trend in evolutionary thinking attempts a new approach. It squarely confronts the three facts that are widely held to be self-evident: (1) we evolved on the savannah; (2) as Darwin revealed, evolution takes place through natural selection; (3) it is proving very difficult to demonstrate that our bodies are adapted by natural selection for life on the savannah.

For a century or so this diabolic triangle has been tackled by battering away at proposition number three, and saying it is only a matter of time before we understand. The new thinking has elected to batter away at number two. Perhaps, it suggests, we are wasting our time looking for an adaptationist solution; perhaps there simply isn't one. Perhaps evolution does not take place only through natural selection; there may be other forces at work which Darwin failed to identify.

That might be one way of cutting the Gordian knot. All that remains is to find out what those forces are, and how they might complement or even negate the laws of natural selection, allowing it to be suspended or overridden or outflanked; that would enable chance to take a hand in the game, so the argument runs.

Suppose a genetic mutation arose by chance . . . Suppose it happened in a small, isolated population where it could quickly become established . . . Suppose it was not an ordinary run-of-the-mill gene (affecting, say, eye colour) which mutated, but a master gene which triggers or controls the operation of other genes, or affects relative rates of growth in different parts of the body . . . In the Burgess Shale, high in the Canadian Rockies, fossils were found of tiny invertebrate creatures living 530 million years ago, so weird and varied that the accepted laws of

Darwinian natural selection seem scarcely to have begun to operate on them. Suppose some other more random factor dictated their designs, and suppose that 530 million years later that other factor momentarily resurfaced and produced arbitrary changes in the ground plan of one particular African primate . . . Or suppose that natural selection only dictates certain key areas of genetic inheritance, while other areas which have escaped our attention are left to chance because they are too marginal to affect survival. Suppose some convergence or concatenation of such changes happened to . . . Suppose . . . or maybe . . . The suppositions have aroused a great deal of interest, but so far none of them has crystallised into a convincing explanation of the emergence of man. Arguably, the law of natural selection, like the law of England, does not concern itself with trifles, and small physical variations may arise within any species (like the exact number of stripes on a zebra) which do not affect its chances of survival one way or the other. But the differences between men and apes are not of that order. It is conceivable that by genetic chance a gorilla might be born hairless, just as on very rare occasions a human baby is born hairy. But hairlessness would not become established in a band of gorillas — however small and isolated — because the nude individuals would be the least likely to survive and procreate in the type of environment where they live. Chance proposes; natural selection disposes.

When Einstein was asked to express in a few words how the idea of relativity came to him he replied, 'I ignored an axiom.'

It is time to tackle the problem of the diabolic triangle by ignoring an axiom. Eliminate proposition number one, that man evolved on the savannah. Postulate, just for the sake of argument, that the crucial early stage of humanoid evolution did *not* take place on the savannah.

We are now confronted with a brand new question: Where *did* it take place?

2

Fossil-Hunters

'Practically all palaeontological discoveries can
be described as bones of contention.'
                                                John Napier

There are two ways of trying to reconstruct the evolutionary history of a living species. One way is to examine
extant specimens, studying their behaviour when they are
alive and their anatomy when they are dead. The other
way is to hunt for the fossilised bones of their distant
ancestors.

These approaches are complementary, each telling us
things that the other cannot. The fossil-hunters' contribution will be considered first, because they loom largest
in the public mind as the evolutionary experts. Whenever
a journalist needs to consult an expert on human origins,
or a television producer wants to make a programme on
the subject, these are the people they approach.

There are several good reasons for this. Hunting for
pre-human fossils calls for a rare combination of physical
and mental abilities, such as courage, stamina, determination and a fair amount of luck, so those who succeed
in it tend to be energetic, eloquent and extrovert. Frequently their discoveries make news, and almost as frequently the claims they make about their discoveries result
in noisy disputations — which further endears them to
the media.

They have great expertise in the comparative skeletal anatomy of humans, apes and the prehistoric hominids whose bones they unearth. Because the skeleton is the only part of the body which fossilises, their knowledge of the non-skeletal organs is much hazier, and when questioned on matters outside their own province they have occasionally been known to give silly answers.

By contrast, anatomical research was long regarded as the plodding and non-glamorous end of evolutionary investigation. Ernst Mayr, Harvard's most authoritative analyst of the processes of biological evolution, once pointed out: 'Ideas based on a study of comparative anatomy have in no case been refuted by subsequent discoveries in the fossil record.' Despite this, anatomists can produce nothing which makes quite the same impact on the public imagination as the empty eye sockets of an ancestral skull.

The most informative and enlightened account of the fossil-hunters' search for human origins is Roger Lewin's *Bones of Contention*, published in 1987. The story begins early in this century, when the race was on to discover the 'missing link' between man and the common ancestor he shared with the apes. With the whole world to choose from, there was considerable uncertainty about where to start digging.

Darwin had predicted that the birthplace of our species would be found in Africa, but for a long time there was a peculiar reluctance, even among Darwinians, to follow up that lead. They were not comfortable with it; they were seeking the origins of the lords of creation, and that conjured up a specific image in their minds. We cannot afford to be patronising about this. Even today, when an illustrator is asked to draw a progressive line of creatures beginning with an ape, growing steadily more erect and intelligent, and ending with a human being, you can be fairly certain that the human being at the end of the line will be male, adult, and white.

Fossil-hunters began digging in Europe, and their hopes were high. The first of the Neanderthal fossils had already been discovered there within Darwin's lifetime. Neanderthals failed to qualify as the missing link because no one could believe we were descended from such ugly brutes, with prognathous faces and villainously low foreheads — even though their brain-cases were slightly larger in capacity than our own. But they reinforced the idea that Europe was good fossil-hunting territory.

Thus, when some practical joker planted the cranium of a modern man and the jaw of an orang-utan in Sussex in 1912, the spuriousness of this missing link — 'the Piltdown Man' — was not finally established until forty years later. The victims of the hoax, as one of them later confessed, were only too ready to believe that 'the first man was an Englishman'. In the '20s a counterbid was made on behalf of the United States. A tooth dug up in 1922 was hailed by H. F. Osborn of the American Museum of Natural History as a Pliocene relic of 'Nebraska Man'. (It was later identified as the tooth of a peccary, a species of pig.)

By the 1920s the prevalent belief was that man had emerged in Asia. The first hint of this had come in the 1890s, when Eugène Dubois dug up, in Java, the first specimen of the species now known as *Homo erectus*. Asia was acceptable where Africa was not, and there was no lack of theoretical arguments to buttress the belief in our Asian origin.

It was reasoned that the enervating climate of tropical Africa was the reason why the African apes had been too languid to scale the evolutionary heights; our own ancestors must have been the product of some more bracing clime. And Asia was, after all, the site of very old civilisations, which suggested that man had been around there longer, and gained a head start. So the experts came down firmly on the side of Asia, and the question was regarded as settled.

The man who brought Africa back into the picture was Raymond Dart. He was an Australian who had studied in London, and he applied for a Chair of Anatomy in a South African University as a way of furthering his career. He did not go there with any presentiment that it was good fossil-hunting country. His mentors had assured him that it was not, and he saw no reason to doubt them.

Yet within two years, in January 1925, he announced that he had made an important discovery. He had been given a small fossilised skull found by limestone workers, plus an endocast of the skull — limestone that had solidified inside the cranium, producing an exact replica of the size and shape of the brain. It later became famous as the 'Taung baby'; the teeth attested that it had died young.

The face was ape-like, but Dart detected some features which struck him as more human than ape-like. One of these concerned the hole in the base of the skull (*foramen magnum*) which indicates the angle at which the skull was set on top of the spine. From its position Dart deduced that the creature belonged to a species which had walked upright. He was excited by it and acted more precipitately than scientific protocol dictates. He sent a letter to *Nature* (published in February 1925) which constituted, in effect, a claim to have found the missing link.

With the exception of a Scottish doctor called Robert Broom, also working in South Africa, nobody believed Dart. He was denounced and derided for not recognising the skull of a young chimpanzee when he saw one, for not realising the brain was far too small for any ancestor of humans, for imagining the missing link could possibly turn up in South Africa. He was even attacked for classical illiteracy in the name he had chosen for his specimen — *Australopithecus*. ('Australopithecus' means 'Southern Ape', but the word comes half from Greek and half from Latin.) Sir Arthur Keith, the dominant figure at the time in this branch of science, examined a plaster cast of the

Taung skull and gave his opinion on Dart's interpretation of it. 'These claims,' he declared, 'are preposterous.'

Dart's ideas began to sink into oblivion. New books on human evolution made no mention of his South African discovery. In 1931 he submitted to the Royal Society a monograph intended as the definitive document on the skull and its significance. The Royal Society rejected it, and it has never been published.

Eleven years after the Taung discovery, Broom found another Australopithecine, this time an adult. Yet it was not until eleven years later still, in 1947, that the tide of professional opinion began to turn in Dart's favour. A spate of recantations followed, led by W. le Gros Clark and crowned by Sir Arthur Keith's succinct admission: 'Professor Dart was right and I was wrong.'

On his ninety-second birthday Dart reminisced: 'I knew people wouldn't believe me. I wasn't in a hurry.' Anyone putting forward an idea which strikes the scientific establishment as 'preposterous' cannot expect to make converts overnight, and 22 years struck Dart as being about par for the course.

From that point on, hunters for the missing link abandoned their dreams of visiting China and focused on Africa instead. In the 1960s and 1970s a series of remarkable discoveries were made — by Louis and Mary Leakey in Tanganyika, by their son Richard in Kenya, and by Don Johanson and Tim White in Ethiopia. There was often fierce disagreement over where the various fossils fitted into the family tree of man's ancestry, or whether they had any place in it at all.

But one general trend appeared to emerge. The hunt was moving further and further north, and the fossils were getting older. The possibility arose that South Africa was not man's original birthplace after all. Taung's tribe of 'southern apes' may have been the descendants of a species of northern apes which migrated south along the Rift Valley.

But Dart's find had come first. It held the field long enough to give rise to an imaginative concept of how man's earliest ancestors may have lived. The fossils found in South Africa by Dart and Broom had been found in a hot, dry area, buried in cave debris, mingled with horns and baboon skulls and other indications of a grassland habitat. Perhaps man was the ape who moved out onto the savannah . . . ?

This theory proved instantly acceptable as an answer to the perennial question of how man originated. In the public mind the images inspired by Kipling's *The Jungle Book* and Rice Burrough's *Tarzan of the Apes* gave place to a vision of our earliest predecessors striding across the grassy plains which cover half the total area of the African continent.

Once such a concept takes hold it becomes extremely hard to dislodge. Scientists use it, tentatively at first, as a model. They construct hypotheses around it, publish papers on it, commit themselves, acquire a vested interest in its continued acceptance. It becomes the conventional wisdom. Savannah explanations were advanced to account for hairlessness, bipedalism, sexual bonding, tool using, and a whole spectrum of other human features.

If the Ethiopian discoveries had been made before the South African ones, it is probable that none of that would have happened. The Hadar site, in the Afar peninsula, is now arid, but there are geological deposits there which prove that it was once a lakeside or riverine habitat. The same thing is true of the Olduvai Gorge.

Of all the fossil sites in the Rift Valley, nearly all were wetter when the hominids were there than they are today. (A possible exception is Laetoli, where a couple of hominids left their footprints on a layer of newly deposited volcanic ash.)

All the other Rift Valley sites, now desert or near-desert, were then green. Don Johanson summed up what is known of them: 'Once they were lush lake regions, swarming with game, laced with winding rivers and thick stands of tropical forests.' Many creatures other than the hominids left their bones at these sites. They include early ancestors of modern pigs and elephants and herd animals and rodents and horses. They also include species of crocodiles, fish, hippopotamuses, frogs, swamp snails, waterfowl, and the fossilised pollen of rushes and other water plants. The most famous of all the fossils to date is known as 'Lucy'. Her almost complete skeleton was discovered in East Africa among the remains of crocodile and turtle eggs and crab claws. And in 1985, when researchers revisited the Taung site, they concluded that there, too, when the baby died the climate was humid rather than arid as it is today.

So if the prospecting had started in the north and worked down, popular illustrations of groups of *Australopithecus* would have shown them reclining under a shady tree at the water's edge, living perhaps on fruit and greenery and fish. Instead, they are depicted as shaggy creatures trekking through parched grass and a scatter of stunted thorn bushes, turning to scavenging and hunting to supplement their diet.

There has been no retreat from the savannah theory. Lucy has been assimilated into the savannah hypothesis. It is argued that the Rift Valley relics do not represent a cross-section of the indigenous fauna at the time but are merely 'death assemblages', and that although Lucy died at the lake edge, that does not prove she lived there. The fossils must include the bones of swamp and lake species which lived on the spot, and the bones of savannah species which merely visited the place to drink, and died before they could go away again. We have no way of knowing to which category the early hominids belonged.

Savannah theorists would argue that for every hominid that left its bones in the lakeside mud, ten or twenty may have left theirs on the arid plains. They say we should not be too surprised that no evidence of this can be produced because the corpses would have been devoured by carnivores and in that kind of environment the bones would not be preserved. It is a good argument; though it is not easy to understand why, if Lucy and her kind could so easily return to the lush places to die, they would not have chosen also to live in them.

Meanwhile, the fossil-hunters' world had again been shaken up — as in Raymond Dart's time — by a 'preposterous' suggestion. It generated a lot of heat, not only because it ran contrary to their own beliefs, but also because it came from outsiders who had never sweated on a fossil site or pondered over the cusps of a prehistoric molar. It came from a new breed of anatomists armed with a shining new investigative technique — biochemical analysis — and it concerned the dating of the emergence of man.

There has never been agreement between the experts concerning the exact shape of the anthropoid family tree. Louis Leakey, for example, never accepted that any of the *Australopithecus* fossils from any of the sites belonged on the line leading to man. They all belonged, he believed, to side branches which had later become extinct, and man's first true ancestor — like the Holy Grail — was still to be sought for. Others were quite ready to believe that Lucy or some other of the Hadar specimens might have been ancestral to *Homo sapiens*.

What none of them knew with any certainty was how long ago the split between apes and man had taken place. It was generally assumed that it happened a long time ago. People are very different from apes, and evolution is a slow process. Throughout the '60s and into the '70s estimates of up to 30 or 50 million years ago were being put forward as possible dates for the ape/man split, and nobody raised an eyebrow. 'At least fifteen million' was the bottom line.

Suddenly molecular biologists were publishing their own figures. They asserted that *only five million* years had passed since the apes and ourselves had a common ancestor. They worked it out by comparing proteins and nucleic acids from the bodies of living humans and of living African apes, and measuring the differences between them on the assumption that the longer two species have been evolving separately, the greater the differences will be. Between man and chimpanzee the difference turned out to be very small — only one per cent.

The 'five million year' claim was published by Vincent Sarich and Allan Wilson at the end of 1967, and it met with a mixed reaction from the palaeontologists. Most of them, if they heard about it at that time, ignored it. That is always the easiest and most time-honoured response, though not terribly constructive. A few of them flatly stated that it had to be nonsense because it disagreed with their own reading of the fossil record. Some hedged their bets by quietly revising their previous estimates of the divergence dates. For example, 'not less than fifteen million years ago' became 'not more than fifteen million'.

Almost all of them found five million years impossible to swallow. 'They didn't offer any reason why it couldn't be true,' Wilson commented, 'they just felt it was somehow obvious it couldn't be true.' He had come up against exactly the same blank wall as Raymond Dart had done, and was reacting with the same sweet reason, defending his attackers. 'I don't think,' he said, 'one can blame the person particularly because it's the social context they're in.'

But his colleague Sarich did not have a Dartian temperament. He was not about to sit around for twenty years and then say he had not been in a hurry and had not expected to be believed. He expected to be believed, *pronto*. Sherwood Washburn, Professor of Anthropology at the University of California, said of him: '. . . he wanted to convert people faster than I think it was reasonable they should be converted.' He was not interested in arguments about old skulls and teeth and jawbones, saying it didn't matter what the fossils looked like. The old estimates should be abandoned at once because '. . . one no longer has the option' of clinging to them.

For a time there was bitter controversy. The new technique was first attacked as unreliable, then examined more attentively. The turning point came when two other scientists, Charles Sibley and Jon Ahlquist, applied a technique called DNA hybridisation, measuring the differences in the overall structure of the genetic material instead of a detailed nucleotide sequence. They arrived at a time-scale longer than that of Sarich and Wilson, but not much longer — between seven and nine million years, rather than between four and six million.

Conversion took place piecemeal. There was a tendency to take the new evidence on board but to keep it in quarantine by expressing it in the subjunctive, on the lines of: 'If we are to believe the molecular biologists, we should have to conclude that . . .' The nearest thing to Keith's 'Professor Dart was right and I was wrong' came in a paper by David Pilbeam. Pilbeam had special reason to be discouraged by the new time-scale, because it shattered a new and promising theory he had been working on, but in 1984 he wrote: 'It is now clear that the molecular record can tell us more about hominid branching patterns than the fossil record does.'

Sarich expressed the same idea from the opposite side of the barricades and in somewhat patronising terms. He said: 'One doesn't want to be foolish and say that the fossil record contributes nothing. But it is true that what it does contribute has been enormously overrated.'

If one did say that the fossil record contributed nothing, one would be talking through one's hat: it is central and vital to our understanding. Molecular biologists contribute to our knowledge of when the split took place, but they cannot tell us precisely where it took place; they cannot tell us in what kind of habitat it took place; and they could certainly never have told us, as the fossil-hunters did, that the hominids were walking on two legs for millions of years before they developed the big brain.

The salient point is that we are now more able than ever before to close in on a probable time and place for the emergence of one key feature found only in human beings — bipedality.

In the last couple of decades the molecular biologists have moved the date of the man/ape split sharply forward in time; the fossil-hunters have moved the date of the first bipedalists steadily backward in time. From both ends the gap has narrowed.

It can now be said with some confidence that at some time between six or seven million years ago and three-and-a-half million years ago, probably somewhere in north-east Africa in the region of the Red Sea, certain anthropoids stood up and began walking erect. Something must have happened which meant that the near-universal mammalian mode of locomotion (walking on four legs) rather suddenly ceased to be efficient for them. They switched to a mode which was not merely different, but unique among mammals. It might be thought that the cause of such a dramatic change is likely itself to have been dramatic.

The orthodox scenario says no. It claims the cause was a gradual dwindling of the forests, so that they could not accommodate so many arboreal animals; some bands of the apes might have dropped out of the branches and ventured out into open country.

'And therefore,' so the story goes, 'they became bipedal.' That is where the logic breaks down. Why should they? No other savannah animal has done so either before or since. It was very far from an easy option. The price for going bipedal — a price we are still paying today — is very high, and has been largely glossed over. In the next chapter an attempt will be made to count the cost.

The story of the bones tells us much about the origins

of man and it also tells us a few things about scientists. With few exceptions, when confronted with a maverick idea, they are confident they can identify whether or not it is preposterous by the gut instinct they have about it. Most of them feel that this absolves them of any obligation to examine it in detail or to give their reasons for rejecting it.

Possibly nine times out of ten, they are right; but sometimes they are wrong. They were wrong in 1912 about Piltdown Man, and in the 1920s about the Taung skull, and in the '70s about the date of our divergence from the ape. They are wrong today in clinging to the savannah theory without reassessing the flimsiness of the evidence in its favour.

3

The Cost of Walking Erect

'For any quadruped to get up on its hind legs
in order to run is an insane thing to do. It's
plain ridiculous.'

Owen Lovejoy

Walking on two legs does not seem by any means a difficult trick to perform when you belong to a species that
has been practising it for a few million years. Unless, that
is, you are fourteen months old and have just tumbled
down for the 65th time; or unless you have broken a leg
and, not having three others to fall back on, are reduced
to hopping or hobbling with a stick; or unless you are
drunk, or suffering from backache, or growing old.

Because people have found it easy, bipedalism did not
strike the early evolutionists as a difficult thing to explain.
When Darwin wrote *The Descent of Man* he did not even
bother to list it in his index. The sequence of events
seemed clear enough at that time. The first development,
it was assumed, was that the ancestral primates acquired
greater dexterity and intelligence. They learned to make
tools and wield weapons, and it is easier to cope with the
weapons, if not the tools, while standing up with both
feet on the ground and both arms free. 'Then,' Darwin
concluded with a cautious double negative, 'I can see no
reason why it should not have been advantageous to the
progenitors of man to have become more and more erect
and bipedal.'

24

He would have seen good reason if he had still been alive when the fossils of Lucy and her companions came to light. They had small brains, and even at sites where their bones were fairly thick on the ground, there was not the slightest trace of evidence of tool making or weapon wielding. This discovery reversed all previous assumptions about the sequence of events. It became clear that bipedalism did not emerge in the wake of other human-like acquirements, to enable them to work better. It came first.

It should not have occasioned quite as much surprise as it did. It was over half a century since Raymond Dart had inferred from his Taung baby's skull that it must have belonged to a bipedal species, it was over thirty years since his claims had in general terms been accepted.

But there is a wide difference between a logical argument in a scientific magazine and the sight of the almost complete skeleton of a bipedal primate. Lucy and the other Hadar fossils — some over three million years old — were the earliest hominids ever discovered. (By contrast, the date of the Taung site has now been assessed at around one million years old.) The first bipedalists were not semihuman creatures. They were animals opting to walk on their hind legs.

It was a costly option for them to take up, and we are still paying the instalments. The mammalian spine evolved over a hundred million years and reached a high degree of efficiency, on the assumption that mammals are creatures with one leg at each corner and that they walk with their spine in a horizontal plane. Under those conditions the blueprint is one that would command the admiration of any professional engineer. The vertebral column is designed on cantilever principles, as a single shallow arch supported by two pairs of movable pillars; the weight of the internal organs is vertically suspended from the arch and evenly distributed along the length of it. Such a mammal resembles a walking bridge.

Our distant ancestors departed from this time-honoured mode of locomotion and converted themselves into walking towers, with a high centre of gravity and a narrow base. There are some other creatures which make use of bipedalism either some of the time or all the time — kangaroos, for example, and ostriches. But they do not proceed with their spines perpendicular; their total body weight is equally and fairly widely distributed around the point where their feet touch the ground. This gives them much greater stability, rather as a tight-rope walker improves his equilibrium by equipping himself with a long balancing pole.

Human vertical locomotion presents more formidable problems, and any sensible engineer confronted with them would insist on starting from scratch. He would probably suggest a spinal column running down the centre of the trunk, with heart, lungs, liver and other organs arranged around it as symmetrically as possible. Supporting ligaments would be attached to the collar bones rather than the spine . . . and so on. Evolution, of course, does not work like that. Every re-adaptation is a process of make-do and mend.

By now, the human skeleton has undergone a notable transformation. The single arch of the spinal column has been abandoned. Babies are still born with it, but they lose it soon after birth in favour of a perfectly straight spine — straighter than it will ever be again. When they learn to sit up, a forward curve develops near the top of the spine, and when they learn to stand they acquire a second forward curve near the base of it. These kinks are essential to prevent the bipedal primate from falling over.

The lower vertebrae have grown bigger to sustain the unprecedented vertical pressure they now have to bear. The pelvic girdle has been moved into a different plane, and the iliac blades on each side of it have spread and flattened into a saucer-like shape to hold the main weight of the intestines. (Otherwise in a bipedal mammal they would be in danger of slumping into the bottom of the body cavity and prolapsing.)

These improvements have made things a lot easier; it is safe to conjecture that the first few million years of bipedalism were the worst. But even today, metaphorically speaking, Nature should still be displaying the sign: 'Reconstruction in Progress. The Management regrets any inconvenience that may be caused.'

Of all the man-hours lost to industry through various forms of illness, the highest percentage derives from our mode of locomotion. Every doctor in general practice must have lost count of the times he has been assured: 'I would be perfectly all right, doctor, if it wasn't for my back.'

The Dresden pathologist Schmorl demonstrated that the spinal column is the first organ in man to 'age' — a process which at least begins as a consequence of the cumulative effects of wear and tear. Pathological changes in the spinal column have been detected in people as young as eighteen years old. In a recent year in Britain, lower back pain occasioned the loss of nineteen million working days. In the United States it has been calculated that 70 per cent of American citizens are affected by it at some time in their lives.

It has been argued that bipedalism was not a very surprising development in a creature as closely allied to the apes as we are, because when an ape moves through the trees by brachiating (hanging from the branches) it is travelling with its spine vertical, as we do. This is sometimes thought to have pre-adapted us for the new mode of locomotion and made it less of a problem.

In fact, as far as the spine is concerned, brachiating is at the opposite end of the spectrum from bipedalism. For the ape, the weight of the body and legs tends to stretch the spine and minimise pressure on the discs of cartilage between the vertebrae. That is why back sufferers are sometimes put into traction or advised to hang from the top of a door to minimise pressure in the same way.

But when a human is standing or walking or running, each of the vertebrae has to sustain the accumulated weight of those above it. The discs between them are flattened from above downwards and expand outwards. This flattening is comparatively slight in an individual disc, but it can add up to an overall shortening of almost an inch. In the course of a day's work, a normally active man or woman decreases in height as the hours go by and the height is restored during sleep when the discs recover their normal shape. In later years the discs become less resilient, so that the flattening process becomes permanent and the height decreases. All people become a little shorter as they grow older, even without the additional problems which can arise through osteoporosis — the decalcification of the actual vertebrae.

Lower back trouble arises because the kink in the lumbar region of the spine makes it structurally weak and unstable. If extra strain is imposed on it, the lowest vertebra is liable to slip backwards along the slope of the next one up. Such displacements may bring pressure on the nerves emerging from the spinal column, giving rise to pain which may be eased or cured by rest, but is liable to recur.

The discs themselves are encased in a tough outer coating which is not easily ruptured, but this can occur. It happens because in the lumbar region the vertebrae do not sit neatly one on top of the other: they meet at an angle, so that the back of the disc which separates them may be subjected at that particular point to a pressure equivalent to a couple of tons to the square inch. (That alarming figure is the kind of calculation that used to be published to explain the damage done to dance floors by stiletto heels.) Occasionally the disc wall ruptures under the stress, part of the contents leak out and solidify, and the resultant bulge causes acute problems which in some cases can only be relieved by an operation.

Apart from changes in the skeleton, the muscular anatomy also had to be remodelled. Walking and standing, plus the strenuous manoeuvre of rising to an erect posture from sitting or squatting, made heavy demands on the muscles of the legs and buttocks which consequently grew in size and strength. The mass of each leg now constitutes one-sixth of the entire body. The newly developed large buttock muscle (the *gluteus maximus*) needed a new point of anchorage on the skeleton, so a special extension of the pelvic flange evolved to accommodate it.

There was an even trickier problem, to which natural selection has not as yet perfected a solution. That concerned the question of holding the internal organs in place. Above the waist there was no difficulty. In a mammal's chest, the heart and lungs are neatly packed inside a limited space enclosed by the ribs, and when we stand upright they are held up by a powerful transverse sheet of muscle — the diaphragm, already present in quadrupeds.

However, there is no cage of ribs protecting the abdomen. In primitive vertebrates there were ribs attached to all the vertebrae. This is still true of some reptiles, but not mammals — possibly because in the females the abdomen has to be extensible to allow for pregnancy.

In the abdomen, therefore, the viscera — which include in our case about 25 feet of coiled intestinal tubing — are much more loosely arranged. In a quadruped, such as a horse, the force of gravity causes the entrails to subside into the curve of the belly, where their weight (plus, when necessary, the weight of the foetus) is supported by a big, broad ligament attached to the vertebral arch of the spine.

When we began to make a habit of walking upright, the pull of gravity was in a different direction, towards the hind end of the body, and the ligament was ineffective because it was in the wrong plane.

Drastic evolutionary measures have been taken by way of damage limitation, and in humans the wall of the lower belly is now protected by three sheets of muscle criss-crossing over one another like a bandage applied to an awkwardly placed wound. But the reconstruction is not yet complete. There is a triangular section on each side of the lower part of the abdominal wall which is left unprotected by the bands of muscle, so that a sudden exertion — in some cases even a violent fit of coughing — may cause a piece of the small intestine to breach the wall and result in inguinal hernia.

Also subject to the pull of gravity is the blood in our veins and arteries. Standing on the head brings this home to us because we are conscious of the blood accumulating in the head and face. But we are rarely conscious that whenever we are on our feet the blood is pooling in our legs in exactly the same way. We may be aware of it on special occasions, such as getting up after prolonged bed rest. The sudden descent of the blood into the lower extremities can cause dizziness or faintness, and the pa-tient may be advised to lie down, or put the head between the legs so that gravity will restore an adequate blood supply to the brain.

In all mammals the blood travels out from the heart through the arteries to all parts of the body and returns to the heart through the veins. In most quadrupeds most of this blood flow is along roughly horizontal channels because the body is horizontal; it is only in the legs that the blood returning through the veins has to climb per-pendicularly against the pull of gravity.

To achieve this it relies on valves in the veins, which allow the blood to pass through in one direction only (towards the heart), and close to prevent the blood sliding back down again. The valves are therefore much more abundant in the limbs than in the other parts of the body.

Bipedalism imposes extra demands on this system, which it is even yet not fully equipped to bear. Our verti-cal posture means that the heart is twice as high above the ground as it would be if we were quadrupedal, and throughout most of our body the blood's return journey to the heart is uphill all the way. The simple act of rising to our feet, by causing the blood to drain away from the heart, can cause cardiac output to drop by 20 per cent.

The greatest strain is on the veins of the legs. As with the lumbar vertebrae, they are under stress because they are at the bottom of the heap. Sometimes a valve gives way under the pressure. This results in twice the weight of blood pressing on the next pair down, which may also give way. This is the cause of varicose veins, where the accumulated pressure causes the walls of the veins to bulge outwards.

Varicose veins are especially likely to occur in pregnant women. One reason is that, because of our erect posture, the weight of the foetus presses down on the large blood vessels in the pelvis, causing a tendency to increased venous pressure, slowing of the rate of flow, varicose veins and swelling of the legs.

When varicose veins occur in the rectum or anus the condition is known as haemorrhoids or piles. The bulging and bleeding may be more frequent than in the legs because the veins in this area are not equipped with valves, since in the average quadruped they are higher than the heart, and the valves are not needed.

It is not surprising that erect posture has knock-on effects on the skeleton and the bloodstream. Its effects on the hormones, however, take a bit more explaining.

Endocrine glands sited above the kidneys produce hormones which respond to actual or potential physical 'emergencies'. The best known is adrenalin (epinephrine), the 'fight-or-flight' hormone. In situations arousing fear or anger it causes sugar to be released into the bloodstream supplying instant energy, plus clotting agents in case the situation leads to violence and bloodshed.

A hormone which responds to 'emergencies' of a different kind is aldosterone. Its function is to regulate blood pressure, and to inhibit the excretion of salt. The 'emergencies' which stimulate the production of aldosterone are listed in medical textbooks as:

1   surgery;
2   anxiety;
3   a diet deficiency in salt;
4   haemorrhage;
5   standing up.

The first four items apply to all mammals and are readily understandable. If there is a deficiency of salt in the body tissues, aldosterone conserves salt by preventing its excretion into the urine. Loss of blood due to surgery or haemorrhage means that the blood volume must be restored; the action of aldosterone increases the fluid volume of the blood. As for acute anxiety, some of the physical manifestations (sweating, vomiting, loosening of the bowels) are notorious squanderers of salt reserves. An increase in aldosterone production will guard against all these contingencies, since this hormone can inhibit salt excretion in the sweat and the faeces as well as in urine.

The fifth 'emergency' (standing up) applies particularly to bipeds. Rising from bed or from a chair produces a six-fold increase in the amount of aldosterone in the blood. This bears no relation to the amount of exertion needed to attain or sustain erect posture. Experiments have been conducted where a volunteer is strapped to a pivoting board which can be tilted on its axis, thus raising the body to the perpendicular with no physical effort involved. The production of aldosterone still rises just as steeply.

The explanation once again lies in the pull of gravity on the bloodstream. When we stand up, the blood tends to drain away from the head and the heart and pool in the lower limbs. It so happens that the main baroceptors which monitor changes in blood pressure are situated in the neck. For a quadruped this is a perfectly appropriate site; blood pressure at this point will be fairly representative of the pressure throughout the body; when these baroceptors register a change in blood pressure, they trigger appropriate responses such as increased output of aldosterone.

But the receptors are incapable of distinguishing between the 20 per cent drop in pressure which would result from a massive haemorrhage in a quadruped, and the 20 per cent drop in the head and neck region which signals in a biped the act of standing up. The aldosterone levels respond in the same manner to standing up as to surgery. Salt excretion is temporarily inhibited, and total blood volume is increased until it reaches a level which will satisfy the baroceptors that all is well — that is, a level which is adequate for the head and chest and over-abundant for the legs. Once that point is reached, aldosterone production ceases, and the blood volume stabilises at the higher level appropriate to the bipedal posture. It is a system which works smoothly and is in general benign. If we could not attain this instant boost to the blood volume level, a feeling of faintness on rising would be much more common and would last longer.

It does, however, mean that our endocrine glands have a lot more work to do. (To a lesser degree, standing up also increases the output of adrenalin.) A quadruped in a secure environment may spend weeks at a time stress free and with no need for these glands to be activated until the rutting season comes round. But in a normal human being the blood volume and the hormones in control of it zoom up and down like a yo-yo many times in a day's work. In patients who are subject to hypertension or circulatory disorders, this side-effect of bipedalism does not help. It is not what the doctor would have ordered if he had any choice in the matter.

The purpose of underlining the debit side of bipedalism at this length is not to whine about it. If these things are part of the cost of being human, few people would consider the price too high.

The evolutionary point is this: the first human ancestors to walk erect were still *animals*, in the popular as well as the biological sense. The compensating advantages which have since accrued to us were not available to them. The disadvantages of walking erect in the savannah would have been instant, and infinitely more uncomfortable for the beginners than they are today.

From the evidence of the skeletons and the footprints, standing and walking cannot have been as easy for the early Australopithecines as it is for us, even though modifications in the skeleton are interpreted as meaning that they had been practising bipedalism for a considerable time. Lucy's feet, for example, were relatively broader and larger than ours (35 per cent of leg length instead of 26 per cent), and this would have given her a different gait, described by Roger Lewin as '. . . not quite as bad as trying to walk on dry land wearing swimming flippers, but in the same direction'. According to the French palae-ontologists Christine Tardieu and Yves Coppens, the automatic knee-locking mechanism found in modern man was not fully developed in Lucy, so that even standing still would not be quite as easy for her. To appreciate the difference it would make, we should have to imagine standing for any length of time in a situation where the locking mechanism could not easily come into play — for example, in a room where the ceiling was just a couple of inches too low.

For absolute beginners the difficulties were very much greater. This is what the orthodox theory asks us to believe: that millions of years ago a population of apes on the savannah chose to walk on two limbs, instead of running rapidly and easily on four like a baboon or a chimpanzee. They stood up, with their unmodified pelves, their inappropriate single-arched spines, their absurdly under-muscled thighs and buttocks, and their heads stuck on at the wrong angle, and they doggedly shuffled along on the sides of their long-toed, ill-adapted feet.

They would not have done this in the blind faith that after the first few hundred thousand years the process would become easier and might in some way pay off. The incentive must have been immediate and powerful.

The next chapter examines some of the theories about what that incentive may have been.

4

Explaining Bipedalism

'We have to admit to being baffled about the
origin of upright walking. Probably our thinking
is being constrained by preconceived notions.'
Sherwood Washburn and Roger Lewin

One factor contributing to the bafflement is the lack of a
parallel to human bipedalism. Other modes of mammalian
locomotion, such as hopping, gliding, swimming, burrowing through the earth or flying through the air, have
evolved over and over again, in separate species often
totally unrelated to one another. But man is the only
extant bipedal mammal. For purposes of comparison
scientists have to fall back on contemplating mammals
that are bipedal for at least part of the time, and see what
advantage they gain from it.

One promising model of bipedalism was thought to be
the patas monkey. Its home is on the open savannah, and
like many of the small mammals of plains and grasslands,
such as the American gopher or the African meerkat, it
has acquired the habit of standing upright to peer into
the distance and detect the approach of danger. The idea
that man had the same incentive was a favoured theory
at the time when David Attenborough produced his TV
epic *Life on Earth*. For the ape-men, he reasoned, 'a permanent upright posture must have been very useful. The
ability to stand upright and look around might make the
difference between life and death.'

The only thing wrong with that argument is the word 'permanent'. Every one of these horizon-scanning mammals, as soon as it spies the approach of danger, takes flight. It does not stay upright, or retreat backwards on two legs so as to keep the enemy under observation. It dives into a burrow, or shins up a tree if there is one handy. Otherwise it turns and runs away at top speed. And for a savannah ape, as for the patas monkey, top speed would certainly have meant using all four limbs to cover the ground.

Other theories about bipedalism which have surfaced during the last couple of decades have drawn their inspiration from (1) meat, (2) seeds, (3) sex, and (4) sunshine.

The oldest — the proposition that man stood upright in order to kill animals — dates back to Raymond Dart, who published a paper in 1953 called 'The Predatory Transition from Ape to Man'. He was impressed by the fact that in the cave at Makapansgat where a hominid fossil had been discovered there were numerous bones of other animals, including crushed baboon skulls. He took these to be the prey that the hominids had carried back to their lair.

It seemed reasonable enough to suppose that whereas an ancestral forest ape, surrounded on all sides by its chosen food, would remain strictly vegetarian, a savannah ape would be driven by the scarcity of lush vegetation to hunt animals. A whole mystique grew up around the idea that the human race ('Man the Hunter') owes its origin to the lust for red meat. When a small piece of skull from a young Australopithecine was seen to have been pierced twice by some sharp instrument, murder was added to meat-eating as one of the factors in hominid evolution.

While some of Dart's ideas were vindicated, the predatory thesis has not stood the test of time. It is now thought more likely that the hominid in the cave was not a predator, but one of the victims of a carnivore. The two holes in the hominid's skull exactly fit the size and distance apart of the lower canines of a leopard. Later, a detailed survey was carried out over some years comparing the behaviour in the wild of a band of savannah-dwelling chimpanzees with another population of chimps dwelling deep in the forest. The forest chimps ate more meat than the savannah ones, and were noticeably more skilled and co-operative in hunting and killing.

But it was above all the discovery of Lucy which cut the ground from under the hunting hypothesis. The fossil remains at Hadar proved that bipedalism had become a specialised form of locomotion long before the hominids evolved big brains or the use of tools and weapons for chasing and killing big game.

The picture of the ape-man as big game hunter began to be modified. Perhaps, it was argued, he fed off small game, or perhaps he scavenged the remnants from the carcases of animals slain by the big cats. Indeed, on closer inspection, the fossil teeth of *Australopithecus* did not look like the teeth of a carnivore.

In 1970 Clifford Jolly made a move towards breaking the link between meat and bipedalism when he published a paper entitled 'The Seed Eaters'. He saw that the ape-men's molars were flatter than a chimpanzee's, and suggested that they were serving as millstones to crush small objects such as seeds. Long, sharp canine teeth would have been invaluable to a carnivore, even a scavenger, in tearing flesh from bone, but in a seed eater they would impede the sideways grinding motion essential to the crushing process. Jolly suggested that this was why the canines were smaller in the hominids than in the other African apes.

To explain bipedalism, he pointed to the feeding posture of the gelada baboon which has been (somewhat unconvincingly) described as bipedal. Geladas live on the savannah and feed mainly on grass. Grass eaters have to spend many hours a day feeding because grass is not very nourishing, so in moving from one patch of grass to the next geladas do not waste time in walking on all fours. While both hands are still occupied plucking grass and carrying it to their mouths, they shuffle along in a squatting position. Jolly reasoned that picking up seeds would make similar demands on a busy pair of hands, so that a seed-eating ape might adopt the same strategy.

The vegetarian approach aroused a great deal of interest, but the gelada model was unconvincing because when a gelada moves in this way it is only erect from the pelvis upwards. Many primates move with their torsos erect, from the brachiating gibbon to the leap-and-cling lemurs of Madagascar. It has never induced any of them to develop habitual bipedalism. Besides, bipedalism strictly implies only two points of contact with the ground, and a closer inspection showed that geladas for much of their feeding time have three points of contact with it. They are shuffling along on their bottoms.

The search for a model was getting nowhere. Almost any primate can be trained with patience and bribery to walk bipedally over short distances, and their respective gaits were being analysed with the aid of high technology, using filmed images and electromyographs to measure activity in the various muscles. Some people were arguing that chimpanzees and gorillas in the wild often rise up when they are excited and charge along on two legs, and therefore the only difference is that they practise bipedalism incidentally whereas we practise it habitually. It looked like a possible solution. Once you have reduced a difference to 'only a matter of degree' it is a short step to assuming that the difference is unreal and there is no need to explain it.

The locomotor analysts would have none of this. Jack Prost of the University of Illinois at Chicago, comparing chimp bipedalism with ours, reported, 'The motions of their limbs and the forces around their joints are so different that there is a question whether the two activities ought to be given the same name . . . The two are quite distinct and only the similarity of names can be offered to argue that one is a primitive form of the other.' M. H. Day, anatomist at Guy's Hospital Medical School, London, concluded that hopes of solving the problem must await further fossil discoveries, 'with known extant locomotor models looking less and less likely to fill the bill'.

The next new approach to the problem was made by C. Owen Lovejoy in 1981, in a paper entitled ' The Origin of Man'. He did not allow himself to get bogged down in the debates over whether the hominid's food was animal or vegetable. That was irrelevant to his theory. The key question lay in what they did with the food when they found it. In the case of most primates, what they do with food when they find it is to eat it. But there are exceptions.

Sometimes a monkey or a chimpanzee will pick up the food and carry it to a place where it can be eaten undisturbed — especially if she is female and fears it is going to be taken away from her. Sometimes a Japanese crab-eating macaque, if the food is left on the seashore, will carry it down to the water and wash off the sand before eating it. And if the food is rather bulky or awkwardly shaped, it may require two hands to carry it and the animal walks on two legs.

These incidents are far too brief and rare to affect a species' behaviour in general or its evolution. The only reason an animal in the wild transports food regularly and over fairly long distances is in order to give it to another animal. Ospreys carry fish to their nestlings; wolves carry meat back to a mate which has produced a litter. Lovejoy felt that the origins of bipedalism might lie less in the diet and more in the pattern of reproduction.

He was the man to whom Don Johanson turned for an expert's opinion on Lucy and the other Hadar fossils, and how they fitted into the story of human origins. As Johanson summed up the situation, 'We had added our bit to the when and where of erect walking. But the why

of it was something else.' Lovejoy's approach to the question was unexpected. As reported by Johanson, when
asked why the hominid and not the apes became bipedal,
he responded:

'Would you like to talk about sex?'

In Lovejoy's thesis as finally presented there was one
new idea and a modified version of an old one.

The new idea was that bipedalism did not evolve on the
savannah, but in the forest. The savannah idea seemed to
him absurd on the grounds that '. . . no hominid could
ever have ventured out on the savannah as a stumbling
imperfect walker and learned to do it better there. If it
had been unfitted for erect strolling on the savannah, it
would not have gone. If it had gone, it would not have
survived the trip.' (And if not the savannah, he automatically assumed, it had to be the forest. After all, where
else was there?)

The old idea was the sex angle — the concept that the
hominid may have been pair-bonded. The most popular
aspect of the hunting hypothesis had been the picture of
a roving male bringing back sustenance to a home-based
female. It made the commuter-and-housewife partnership
seem part of Nature's plan. Since bipedalism came too
early to be caused by hunting, Lovejoy rearranged the
order of events, so that pair-bonding came first. The
ancestral ape became pair-bonded in the forest. Through
carrying handfuls of vegetable food to his family he gradually perfected bipedalism. That made him better qualified
to walk out at a later date onto the savannah where he
eventually learned to add meat to the menu.

That is a somewhat brutally compressed summary of a
thesis set out by Lovejoy in a complex scientific exposition
full of loops and feedbacks. Even if we accept for the
moment that the ancestral apes may conceivably have
become pair-bonded, there are some basic weaknesses in
the scenario.

One is that male primates do not make a practice of

offering food to females — not even when they are pair-bonded. The only pair-bonded ape is the gibbon, and he cements the bond not by offering sustenance but by chasing off any potential rivals, including his own growing sons, that may threaten his monopoly. Lovejoy urges that primate relationships are very varied. He cites the marmoset as an example of a species where the input of paternal child-care makes it possible to rear more offspring (marmosets normally have twins).

But what he does not make clear is that the care consists not of feeding the young, but of minding the babies while the female goes off to forage for herself and then returns to suckle them. The likelihood of a gorilla, for example, bringing food to a female or juvenile is as improbable as a cow bringing grass to a calf and for the same reason: once the young are weaned their food is all around them and they can help themselves.

The other weakness is the unquestioning assumption that an ape on the ground carrying things would automatically do it on two legs. Graham Richards published a paper in 1986 challenging the long-held theory that the transition from trees to the grassland brought with it a sudden freeing of the hands for manipulation and tool-making. He concluded that 'Talking about "freed hands" when there is little indication as to whether they were either enslaved in the first place or in any genuine sense "liberated" subsequently, helps nobody understand what really happened.'

In practice, knuckle-walking chimps and gorillas can free one hand instantly whenever they need to.

In 1985 Russell Tuttle and David Watts, of Chicago University's Department of Anthropology, analysed the result of 1,700 hours of observation of mountain gorillas in Ruanda and Zaire. The animals spent 98 per cent of their time on the ground, most of it sitting or squatting to feed. When they needed to reach up or out for food they used a tripedal posture thirteen times more often than a bipedal one. Out on the savannah, with the need for faster and more sustained locomotion, bipedalism would be even less likely. A chimpanzee running off with a stolen banana runs on three legs and carries it. Stealing a blanket, it runs on three legs and drags it. The types of food-load which would force it to resort to bipedalism are very rare in the wild. An experiment with Japanese macaques shows that when confronted with a pile of potatoes they may grab an armful and hobble away with them on two legs to a quiet spot. But it is not a good model for primates in Africa, least of all on the savannah.

Perhaps the unlikeliest aspect of Lovejoy's thesis is the idea of females and young being left unprotected in the middle of the plain while the male wanders off to forage at a distance. It would be risky and untypical primate behaviour. In the case of baboons there is an elaborate strategy for ensuring that the females and young are always kept in the middle of the troop and whenever they move on they are flanked by male outriders. The hunter theorists were fond of the phrase 'back to the lair', but primates do not dwell in lairs, and even if they wished to adopt the habit, the open savannah is the last place they would be likely to find one without burrowing underground.

The sunshine theory was touched on in 1970 by R. W. Newman, and elaborated in 1984 by Peter Wheeler. Feeding habits had dominated the discussion for some decades, but this theory dropped the topic of food and returned to considering the nature of the terrain.

The last time that approach had been employed was in the horizon-scanning theory modelled on the patas monkey. But when an animal bred in the forest moves out onto the open plain, it may encounter one problem even more urgent than the fear of predators. In tropical Africa it is very much hotter away from the shade of the trees, especially in the middle of the day when the sun is almost vertically overhead. High levels of direct solar radiation can cause considerable stress, so, other things being equal, any method of lessening this stress would be favoured by natural selection.

Wheeler gives figures, derived from experiments with models, showing that at midday in the tropics an ape on four legs exposes seventeen per cent of its total body surface area to the rays of the sun, while an upright hominid exposes only seven per cent, thus absorbing less than half as much heat by direct solar radiation. The seven per cent consists of the top of the hominid's head and shoulders, suggesting that the hominid stood up to keep cool, and scalp hair was retained as a shield over the exposed parts, reflecting and reradiating much of the heat before it could reach the skin.

One difficulty the theory does not address is the fact that an ape has to expend muscular energy — far more than we do — in order to maintain the upright posture, and the expenditure of muscular energy tends to raise body temperature. This fact would certainly diminish, if it did not cancel out, the cooling effect of standing up. Another snag is that erect posture would only work as a partial cooling device while the sun was at or near the highest point of the sky. Seeking to exploit a niche as a noonday forager would be an unrewarding tactic for a mammal allegedly so ill equipped to resist heat stress.

But the greatest weakness is the one that haunts every variant of the savannah hypothesis, namely: if this was such a good solution of a common problem, why was only one species driven to resort to it?

None of the suggested answers is good enough. Wheeler argues that most savannah animals, such as lions and zebras and hyenas, are protected when exposed to heat stress by a mechanism (the carotid rete) which protects the brain from overheating by maintaining it at a lower temperature than the rest of the body. Primates spent most of their evolutionary history in shady places, so they never evolved this protective device.

This explanation does not work because several primate species have at various times made the transition from trees to savannah, including baboons and geladas, patas monkeys and vervet monkeys, and they have all flourish- ed there without being forced into walking on their hind legs. Some escape the midday heat in the same way as many other mammals, by taking a siesta in the shade of a rock or thorn bush; others continue foraging, and they too appear to come to no harm. The most savannah- adapted of primates is the baboon, and its mode of loco- motion has evolved in the opposite direction from bipedal- ism, namely, towards resembling ever more closely the standard quadrupedal gait of ground-dwellers such as the dog.

There remains the water theory. The nearest primate model here is the proboscis monkey, which often resorts to bipedalism not from choice but by necessity. In the award-winning documentary film *Siarau* (Partridge Films, Ltd. 1984), there is remarkably vivid footage of a band of these animals walking along on their hind legs, up to their chests in water. Leading the way in this shot is a female carrying an infant in her arms, the clasp and the posture strongly reminiscent of a woman carrying a baby.

The proboscis monkey has never in the classic sense made the transition from the trees to the ground, because it lives in mangrove trees in the coastal swamps of Borneo. More often than not when it climbs down a tree, what it encounters is water. If the water is deep the monkeys swim. They are excellent swimmers and can swim for miles, and may be seen diving into the water from the tree tops. If the water is shallow they wade.

Occasionally, and especially at low tide, they are seen on relatively dry land — on a mud flat or a sandbank — and their mode of locomotion is sometimes on four legs and sometimes on two. Local fishermen speak of family groups of these monkeys 'going for a walk', and by this they mean walking on two legs. Film of proboscis monkeys on the ground in the wild is rare because of the terrain, but some shots were captured by a Japanese cameraman up to his thighs in squelchy mud. They feature in a documentary film entitled *Long Nose, Long Tail*, directed for Japan Broadcasting Corporation by Yuiche Naka, and they include a sequence in long-shot of a group of proboscis monkeys walking bipedally.

The difference between the bipedalism of the proboscis monkeys and the African apes is striking. Gorillas and chimps make short energetic dashes on two legs, often moving sideways, and accompanied by hoots or squeals of excitement. In the case of gorillas it is invariably part of a display ritual, showing off or threatening, somewhat analogous to a human war dance. (In such a context the human dancer may dance sideways on *one* leg while emitting ritual noises, but this is unlikely to lead to monopedalism in daily life.) By contrast, the wild proboscis monkeys in the Japanese film, having acquired their bipedal gait in water, are seen walking calmly on the ground through the trees in single file.

Inundation of the habitat is their incentive for bipedalism. For proboscis monkeys crossing a stretch of water a couple of feet deep, walking upright offers only one single advantage, but it is an offer they cannot refuse. It enables them to breathe, whereas if they walked on four legs, their heads would be under water.

In all the savannah scenarios the disadvantages of bipedalism — unstable equilibrium, disruption to skeletal, muscular, circulatory and hormonal systems — would be incurred in their most extreme form immediately, and would only ease off in successive generations as the bones gradually changed their shape and the new muscles developed. On the other hand, the supposed advantages would be non-existent or minimal in the first generation while the ape was still an ape. They would only accrue to its distant posterity.

In other words, terrestrial bipedalism would only

become advantageous provided it had been diligently practised for thousands of years while it was still awkward and laborious and the behavioural rewards were infinitesimal or nil.

In the aquatic scenario the position is reversed. Walking erect in flooded terrain was less an option than a necessity. The behavioural reward — being able to walk and breathe at the same time — was instantly available. And most of the disadvantages of bipedalism were cancelled out.

Erect posture imposes no strain on the spine under conditions of head-out immersion in water. There is no added weight on the lumbar vertebrae. The discs are not vertically compressed. (An astronaut in zero gravity gains an inch in height in the first days in space, and immersion in water is the nearest thing to zero gravity on planet Earth.)

In water, walking on two legs incurs no more danger of tripping over and crashing to the ground than walking on four. There is no distension of the veins because immersion prevents the blood from pooling in the lower limbs.

Standing up in water does not trigger secretion of aldosterone, salt retention or higher blood pressure. The reverse is the case: head-out immersion causes a prompt and marked *fall* in systolic and diastolic blood pressure, plus increased excretion of salt in the urine. This effect is so marked that in patients suffering from high blood pressure and sodium imbalance, such as children with nephrotic syndrome and women with late pregnancy toxaemia, head-out immersion is effective in relieving the symptoms. Some women suffer side-effects from the orthodox drug treatment, and in these cases spells of immersion therapy can reduce the amount of drugs they have to take.

Water thus seems to be the only element in which bipedalism for the beginner may have been at the same time compulsory and relatively free of unwelcome physical consequences.

An early objection to the aquatic solution was that primates in general, and apes in particular, generally have an aversion to water, so it is hard to believe that any population of African apes could have chosen to go into the sea. A much more probable explanation is that the apes stayed where they were and the sea came in to them. The vicinity of Hadar was one of the most unstable spots on the surface of the earth at the time of the ape/man split.

More than twenty years before Darwin published *The Origin of Species* the eminent geologist Sir Charles Lyell was working on three volumes of his classic work *Principles of Geology*, published between 1830 and 1835. Lyell was a friend of Darwin's, and his ideas about the history of the earth helped to blaze the trail for acceptance of Darwin's ideas about the evolution of living things.

Lyell described Earth as a planet which was constantly changing, raising mountains where there had been plains, and land where there had been sea, eroding peaks, giving birth to new islands in the oceans, grinding rocks into sand. He argued that these changes were due to natural causes which were still operative and that therefore the world must still be changing. If the changes were imperceptible it was because they took place very slowly. Lyell's geology required his readers to discard the Church's teaching that only a few thousand years had elapsed since the world was created. He opened people's minds to the idea of earth-time stretching far into the past over millions of years of gradual change, and Darwin adopted the same time-scale when he wrote his own book.

It is impossible to understand the processes of the evolution of living things without reference to the geology, geography and climate of the places where they live. Within fairly narrow limits, these factors dictate the kinds of life-forms which can exist on the world's surface, in the thin outer layer of land and water now christened the biosphere. It supplies the chemical elements of which all living things are made, and has determined the range of temperatures within which they can survive.

If the earth's axis were not tilted at an oblique angle to the sun, there would be no oak trees and no dormice — because there would be no seasons to give rise to leaf fall and hibernation. If the earth were not circled by a satellite, there would be no tidal beaches and none of the creatures specialised to exploit them.

In popular accounts of evolution there is a tendency to devalue the dynamic planet as a factor in evolution, and replace it with an inner life force. For example, the great evolutionary landmark 350 million years ago, when the first backboned fishes came onto the land, is often described in terms of aspiration and conquest, as if the fish climbed up the beach as John Hunt climbed Everest — because it was there.

It cannot have happened like that. No advantage could possibly accrue to the species that would even begin to outweigh the pains and perils of the first few hundred generations of being a fish out of water. It happened because the earth is not static. Lakes and pools shrink or dry up; rivers change their courses; land sinks below sea level; estuaries silt up. The first land vertebrates, like the little mud skippers found in the tropics today, were creatures who found themselves stranded for at least part of the time by movements of the tide or by seasonal droughts. They did not move out of their natural habitat to advance upon the land and colonise it. They stayed where they had always been. It was the earth that moved.

Since Lyell's day we have learned things about the movements of the earth that he never dreamed of. He introduced the concept of the earth's fabric on each of the great continents rearing up, contorting, being compressed or pulverised, fractured by ice or molten by volcanic fires. But he thought the continents themselves were firmly anchored in positions they had occupied since the original formation of the earth's crust.

In 1912 Alfred Wegener published two articles outlining the hypothesis of Continental Drift — the idea that continents have moved horizontally relative to one another and to the ocean basins, and that the movement represents the splitting and drifting apart of land masses which had formerly been contiguous. He supplied a mass of evidence based on geological data, measurements of gravity, facts about climate in earlier ages, and the eccentric and unexplained distribution of fossil plants and animals throughout the continents as they are today.

When the physicist W. Lawrence Bragg organised the first British discussion of Wegener's theories in Manchester, he was staggered by the response. 'The local geologists were furious; words cannot describe their utter scorn of anything so ridiculous as this theory.' Wegener's opponents spent little time addressing his evidence, and none at all in attempting to find alternative explanations of the phenomena he described.

They stressed that they could not see how Continental Drift worked — with the implied corollary that if they could not see the method, then the method did not exist. They outlined theoretical proofs that it could not possibly work. Lecturers defending Wegener's ideas were often greeted with the kind of boos and catcalls usually reserved for a third-rate vaudeville act at the Glasgow Empire. Wegener died in 1930 during an expedition to Greenland while this hooligan phase of geological analysis was at its height. He would have had to live another thirty years to see the tide of opinion swing in his favour, and Continental Drift transmuted from heresy to orthodoxy.

The Wegener hypothesis about moving blocks of the earth's crust has particular relevance to the story of man's origins. In north-east Africa there is a unique and extraordinary geological situation where not two but three crustal blocks converge. One line of rifting continues 280 miles to the north to form the Wadi Aqaba-Dead Sea-Jordan rift. Another runs east 600 miles to form the Gulf of Aden. The third runs south along the whole length of the African Rift Valley, marked by signs of continuous volcanic activity lasting over eleven million years.

This convergence of geological forces and tensions has its focus in the district of Ethiopia known as Afar, which gave its name to Lucy's scientific title: *Australopithecus afarensis*. The region has become as much a Mecca for tectonic, volcanological and geological researchers as for fossil-hunters, and they have made equally remarkable discoveries.

As with the fossil-hunters, the information gathered by the geologists has given rise to excitement and controversy in about equal measure. Paul Mohr, the geologist responsible for reviving the old name of Afar for the region, summed up what had been discovered, and the debates over interpretations, in a paper bearing the one-word title 'Afar'.

He made it clear that despite disagreement over some aspects of the new data, there was virtual unanimity on the one point which chiefly concerns us here. He wrote as follows: 'What is more or less agreed upon is that by the late Miocene [seven million years ago] a marine basin had become established over northern Afar . . . and these conditions persisted until the isolation and desiccation of the Salt Plain arm of the sea some 70,000 years ago (CNR-CNRS team, 1973).'

The sea flooded into Afar, but it never flooded out again. Further movements of the jostling crustal blocks sealed off the Sea of Afar from the ocean, and over millions of years the water evaporated. A similar thing happened, at the northern end of the rift, to the Dead Sea in Israel and the end result will be the same — it will become brinier and brinier until nothing is left but a salt plain. The Afar Depression is now one of the hottest and most nearly impassable deserts in the world. The salt deposits are thousands of feet deep. For the tribesmen who sometimes load the salt and take it on the long trek to the cities, it is the only marketable product the territory affords.

At the eastern end of the salt plain is a highland region called the Danakil Alps. Geologically these highlands are described as a 'horst' — a chunk of the earth's crust torn away from one of the crustal blocks and tilting down at the western end. Danakil was never covered by the ancient Afar sea. Geologists can tell, from the rocky debris, precisely how far up the side of it the water rose. Some of the plants on it have become differentiated from their counterparts to the west of the salt plain, as normally happens when an island is separated from a mainland.

These things were happening around the time suggested by molecular biologists as the beginning of the ape/man split. They were happening close to the region where the very earliest of the hominid fossils have been discovered. They took place in latitudes where a great unbroken band of forest formerly extended from the west of Africa across the continent and on throughout most of Asia. The region affected by these geological disturbances would therefore have been the kind of habitat in which apes flourished and still flourish.

It is hard to escape the conclusion that some of the apes must have found their environment suddenly and radically changed. Some may have been cut off from the rest of the population by an arm of the sea as it flowed into the terrain which now lies far below the great Salt Plain. The American Leon P. LaLumiere, of the Naval Research Laboratory in Washington, was the first to suggest Danakil Island (now the Danakil Alps) as a place where one such group may well have been marooned.

Geographic separation of this kind provides ideal conditions for rapid evolutionary change. In 1963 Ernst Mayr, in his classic work on this subject, *Animal Species and Evolution*, regarded it as the commonest cause of species diver-

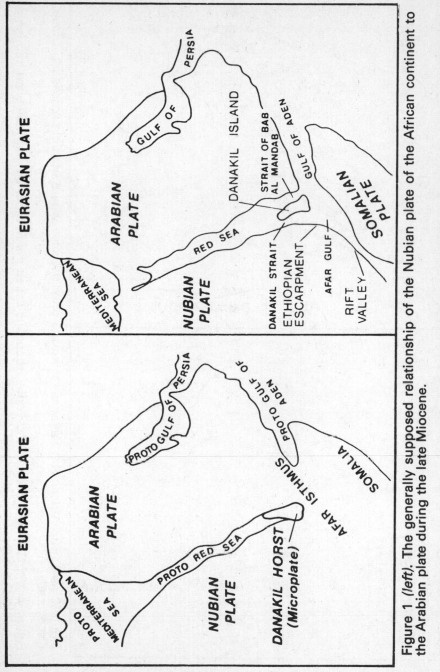

Figure 1 (*left*). The generally supposed relationship of the Nubian plate of the African continent to the Arabian plate during the late Miocene.

Figure 2 (*right*). The configuration of the same region as displayed in Figure 1, but as it may have been at the beginning of the Pliocene.

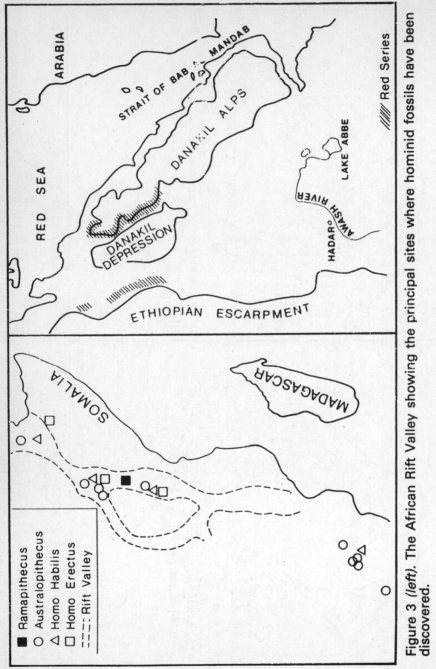

Figure 3 *(left)*. The African Rift Valley showing the principal sites where hominid fossils have been discovered.

Figure 4 *(right)*. The location of the Tertiary deposits called the Red Series that may yield hominoid fossils if this hypothesis is correct.

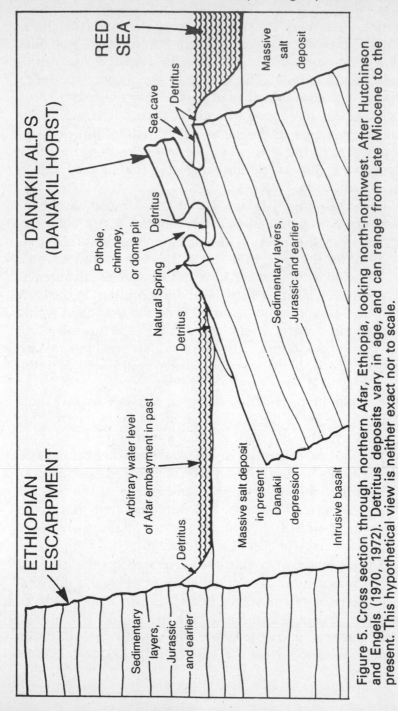

Figure 5. Cross section through northern Afar, Ethiopia, looking north-northwest. After Hutchinson and Engels (1970, 1972). Detritus deposits vary in age, and can range from Late Miocene to the present. This hypothetical view is neither exact nor to scale.

gence. He wrote: 'That geographic speciation is the almost exclusive mode of speciation among animals, and most likely the prevailing mode even in plants, is now quite generally accepted. The theory of geographic speciation is one of the key theories of evolutionary biology.'

An alternative mode sometimes occurs where part of a species remains in the same shared habitat but occupies a different niche in it — 'sympatric' or 'in-the-same-place' speciation. But no one has suggested that as the mode of our own origin.

Savannah theorists agree that there must have been geographic separation, but they suggest that a voluntary move from the trees to the savannah was sufficient to account for the split. In theory it could be so, but it would have taken much longer to result in species divergence. The isolation would be partial and intermittent in the early stages. The apes who ventured onto the grassland would sometimes return to the trees for shade or safety, or encounter individuals who were still arboreal but making a brief sortie. Any such encounter could lead to inter-breeding, and the genes would once again be mingled in the common pool. The savannah version looked more convincing in the old days when fifteen or twenty million years were allowed for a gradual speciation.

The more absolute the isolation, the quicker the change. R. R. Miller in the '50s and '60s researched speciation of fish in the freshwater springs and creeks of the western North American deserts, and demonstrated that speci-ation in small isolated bodies of water might proceed one thousand times as fast as in the ocean.

The aquatic theory has been criticised on the grounds that the ape/man divergence took place so rapidly that '. . . there would not have been time' for a species to go through an aquatic phase during the period of the fossil gap. But time is on the side of the aquatic theory. Really rapid speciation calls for stricter isolation and a more abrupt change in the environment than the savannah theory provides for.

It is impossible to deduce exactly the kind of environment in which the apes lived. Nowadays we publish maps with clear black lines defining the boundaries between land and water, and try to separate the two by draining bogs and fens and turning wetlands into marinas. But the earth is not quite like that, even today. All over the world there are areas like the Everglades of Florida, the Sundarbans of the Ganges delta, the mangrove swamps of the East Indies, and the Okavango in Botswana which are hard to define either as land or water. In the Amazon basin there are tracts of territory — of a total surface area equal to that of Great Britain — which are forested but for more than six months of every year are under water and inhabited by fish and dolphins swimming among the tree tops.

If we wish to know what happens to an ape in an aquatic environment, we can turn to one species whose aquatic credentials are less controversial than those of *Homo*. There was a species — no ancestors of ours and long extinct — called *Oreopithecus*. These apes have been dubbed 'swamp apes' because their bones were laid in mud, which preserved them so well that their fossil remains have been found in large numbers. Some of the skeletons found in Italy were nearly intact. They were once hotly fancied as our earliest ancestors, and one enthusiastic palaeontologist, J. Hürzeler, claimed to be able to trace eighteen resemblances between the skeleton of *Oreopithecus* and that of *Homo*.

Since *Oreopithecus* was expunged from our family tree, interest in the species has waned. It is now recognised that any resemblances between the swamp apes and man are not due to a close genetic relationship. The only other possible explanation of them, if we rule out blind coincidence, is convergent evolution. This is the mechanism by which species inhabiting similar environments and with a similar life-style often grow to resemble one another though they are genetically unrelated and may be separated by half the width of the world.

The idea of convergent evolution providing a link between man and *Oreopithecus* is an illuminating one. Because one of the interesting things about *Oreopithecus* is its pelvis. It is characterised by the short iliac bones which are seen in most aquatic mammals, and in man and his ancestors. The evidence suggests that through living in marshes *Oreopithecus* had become either a good swimmer or a bipedalist — or both.

5

The Cost of a Naked Skin

'There are one hundred and ninety-three living
species of monkeys and apes. One hundred
and ninety-two of them are covered with hair.
The exception is a naked ape self-named *Homo
sapiens*. The zoologist now has to start making
comparisons. Where else is nudity at a premium?'
Desmond Morris

Darwin, as we have seen, believed he could accommodate
human bipedalism within the basic framework of his
theory. He thought it possible that walking upright made
our ancestors fitter to survive. But human nakedness was
a different matter. He had no illusions about that.

In his book *The Origin of Species*, published in 1859, he
gave copious examples of how natural selection worked,
but he did not discuss the human species in relation to
it. After stating his belief that his theory would open the
way to new avenues of future research, he contented
himself with thirteen words on *Homo sapiens*. 'Much light,'
he wrote, 'will be thrown on the origin of man and his
history.'

This reticence is sometimes ascribed to prudence, or
caution, a desire to minimise the hostility he knew he was
going to encounter. If that was his hope it was certainly
not fulfilled, and he cannot seriously have expected it to
be. In the famous Oxford debate of 1860, all the most
heated arguments raged not around the theory in general,

but around the one omitted topic — man's relationship with the apes.

There could have been a different reason for the omission. Perhaps he was not ready to give an account of human evolution while there were things about it he could not understand. He could not, for example, convince himself that being naked could possibly make a primate fitter to survive. This is what he had to say about it:

'The loss of hair is an inconvenience and probably an injury to man, for he is thus exposed to the scorching of the sun and to sudden chills, especially due to wet weather. No one supposes that the nakedness of the skin is any direct advantage to man; his body therefore cannot have been divested of hair through natural selection'.

The most striking thing about that statement is that most modern evolutionists — some of them more Darwinist than Darwin himself — would find it impossible to accept. They start with the axiom that man evolved on the savannah, they observe that he has lost his body hair, and they conclude that nakedness *must*, therefore, have been an efficient adaptation for life in that milieu.

Practically all the contemporary theories about nakedness begin with the concept of a torrid savannah and an overheated ape. That is an oversimplified picture of a savannah environment. It is perfectly true that the days are hot there, sometimes very hot; but in the nights the temperature can sometimes drop as low as 11°C, and the indigenous animals have to live there for twenty-four hours a day.

In 1989 the BBC sent a series of outside broadcast crews to spend an entire day on the savannah, sending back live pictures of the wildlife to be slotted into the viewing programme at intervals during the day. Some magnificent shots of wildebeest, lions and elephants were obtained, but in some ways the most memorable shot was the final one of a couple of broadcasters muffled up to the eyebrows trying to keep their teeth from chattering. As one reviewer summed it up, 'The weather broke, the light went, and Julian Pettifer nearly died of exposure.' It would not have been a good time to explain to Mr Pettifer that his ancestors had shed their fur to make them better fitted to survive in these conditions.

If they had kept their coat of fur they would have been better insulated not only against the cold of the night but also the heat of the day. All desert animals have retained their fur. In hot countries, shaving off even a portion of wool from a sheep's back leads to an immediate rise in its internal temperature and the rate of panting. Desert-dwelling humans like the Bedouin keep their heads and bodies covered when out in the sun, just as northern peoples cover themselves with clothes when out in the wind, in both cases attempting to replace the natural protection that our species has lost.

A covering of either fur or feathers has been the standard equipment for warm-blooded creatures ever since they emerged over a hundred million years ago, and ultimately took over from the reptiles as the dominant form of animal life.

One of the many advantages of such a covering is the ability to adjust the thickness of insulation within seconds. Hairs and feathers have erectile muscles attached to them, so that robins, for instance, can puff out their feathers when the temperature drops; animals can erect their hairs either in response to cold or danger; cats and dogs bristle the hair on their backs when confronted with an enemy in order to look larger and more intimidating.

Humans still retain the capillary muscles in good working order, but their functioning is comically ineffective. Since the vestigial hairs they are attached to often fail to reach the surface of the skin, the only result achieved is goose-bumps. We produce them in cold weather with the subconscious aim of bushing out a non-existent coat of fur. When we are frightened, some primitive level of our brain may try to raise our hackles in the hope of scaring off the enemy.

Hair or fur is a mammal's first line of defence against abrasions — cuts and grazes, the sting of an insect, the scratch of a thorn. Perhaps partly in an attempt to replace this defence, our skin has become thicker than that of our nearest relatives, the apes. Yet at the same time human skin contains more blood vessels and is more richly endowed with nerve endings. This means that even a small lesion is liable to bleed more, and hurt more, than if we had skin like an ape's.

Much in the news recently is another disadvantage of nakedness — the danger to light-skinned races of over-exposure to ultra-violet light and the consequent risk of skin cancer. Most mammals do not incur this risk, except where the hair is missing. In horses, for example, a malignant melanoma may sometimes develop in the bare patch under the tail.

Cancer is the most serious of the possible ill-effects of exposure to ultra-violet rays, but it is not the only one. Exposure to strong sunlight can damage light-coloured skin to the extent of putting the skin glands temporarily out of action. In the heyday of the British Empire, young men and women fresh out of England to serve in India found that part of the white man's burden was susceptibility to an uncomfortable skin condition known as 'prickly heat'.

Since the loss of body hair, human skin has elaborated an auxiliary line of defence against the danger of ultra-violet rays. Scattered throughout the top layer of our skin are small spider-shaped cells (melanocytes) which produce a substance called melanin, after the Greek word for 'black'. The cells respond to ultra-violet exposure by increasing the output of melanin, which is injected into the surrounding skin cells where it forms a protective shield like a minuscule sunshade over each cell membrane on the side nearest to the surface. This darkens the colour of the skin and protects it by absorbing the harmful ultra-violet rays. In the dark-skinned races the protective shield is permanent.

There is some melanin in the skin of apes and monkeys, but its purpose there is more obscure. It is hardly needed as a protection against the sun, for most of these animals live in shady places, and all of them have bodies which are covered with hair. Since their skin colour thus cannot affect their chances of survival, it is not governed by the influence of natural selection, and varies widely between species, and sometimes between individuals. The Celebes ape is born with a black skin which later turns white. The rhesus monkey's skin is pinkish with random splotches of blue. There may be dark-skinned and light-skinned individuals within a single troop of chimpanzees.

In the days when Theology was queen of the sciences, learned men were as curious about their progenitors as we are today. One of the questions they asked themselves was: 'What colour was Adam?' Considering in whose image Adam was created, it was a matter of some delicacy, never satisfactorily resolved. The old monks who had the job of illuminating Holy Writ ignored the issue, and went on drawing little pictures of a white Eve with a white Adam.

Some of the early evolutionists thought they had cracked the problem. Since they could now dispense with Adam, they felt able to jettison the egalitarian Biblical doctrine that all men were brothers: they suggested that the races had always been separate, having evolved independently from the beginning. Behind that hypothesis we seem to detect an unspoken conviction that some of us are obviously descended from a rather better class of ape than others.

We now know that all races share a common ancestry, and there is every reason to believe that our earliest common ancestors lived in Africa, where protection against ultra-violet rays would be at a premium.

This conclusion has been challenged. It is argued that protective pigment could not have evolved by natural selection because the danger it protects against does not usually manifest itself until middle age, when child-bearing is over; therefore, a female who was not protected against skin cancer would produce just as many offspring as a protected one, and her genes would not be selected out. This argument is unsound. Half our genes come from our fathers, and a man's reproductive capacity does not close down as early as a woman's. He can certainly increase the number of his descendants by avoiding late-onset disease and living longer, so that over time natural selection in an African climate would favour dark skin.

Besides, the white races bear the evidence of ancestral pigmentation in their own skins. All human beings, of whatever race, have the same number of melanocytes. In northern races they are largely surplus to requirements and remain relatively inactive because in temperate climes the ultra-violet rays are seldom strong enough to be dangerous. But they are still there, presumably a relic of our heritage.

So the question of racial differences in skin colour does not revolve around the question of why some human populations became black, but why some later faded to lighter shades. The best available theory is that it happened when they migrated from Africa to northern latitudes where the climate was cooler and cloudier and where the food was scarce in winter.

They were then in less danger from strong sunlight and in more danger from lack of vitamin D. A shortage of this vitamin in the diet does not matter too much in hot countries. Cells in the skin can manufacture their own supply, but they cannot do it — any more than a leaf can manufacture chlorophyll — in the absence of sunlight. The best survivors in northern countries would be the ones with the least active melanocytes (that is, the palest skin) because the limited available sunlight could soak into their skin more easily and help to make vitamin D.

So it could be claimed that in the end natural selection did a good job, under difficult circumstances, both for the dark races and the pale ones. As a method of counteracting part of the ill-effects of hair loss, it was indeed a neat piece of damage limitation. The drawback is that it takes a very long time for a human population to evolve the kind of skin colour fitted to a particular latitude. The system is foolproof only as long as people stay put. Nowadays they do not stay put, and sometimes they suffer from it.

For example, when Asian women settle in Britain and continue to live on the same diet that kept them healthy in India, they are sometimes found to be suffering from rickets. A tropical sun is strong enough to penetrate through a thin sari and enable their skin to make vitamin D, but the British climate leaves them deficient in it.

The reverse move — from the temperate zones to tropical ones — can be even more damaging. People of European origin have made their homes in all parts of the world and the possible ill-effects have only recently been fully appreciated.

There has been a startling increase in the frequency with which malignant melanoma occurs throughout the Western world, with a doubling of the frequency of the disease each decade since records started. The rise has been highest among white populations living in hot countries, and Celtic races seem to be especially vulnerable.

It is very rare in dark-skinned people, and the incidence among the whites seems to vary with the distance from the equator. It is high in places like Arizona, Israel, Hawaii and Australia. In New South Wales it accounts for seven per cent of all cancers, as compared with one per cent in the United Kingdom. There are now fears that damage to the ozone layer could accelerate the increase.

'Naked as Nature intended' was a persuasive slogan of the early Naturist movement. But Nature's original intention was that the skin of all primates should be un-naked. Faced with the phenomenon of the bald-bodied ape, she has tried various experiments without finding a universal solution.

Doctors have tried hard to publicise the fact that exposure to the sun is possibly dangerous, and in the long run anti-cosmetic — it is the direct cause of the wrinkles associated with ageing and the broken veins which sometimes appear under the skin. But their campaign has had only limited success. In an American survey, over 50 per cent of those questioned knew that there was a link between sun-bathing and cancer, but two-thirds of that number continued to sun-bathe, and in most cases without using a barrier cream. Westerners cling to their belief that a tanned skin is 'healthier' than a pale one, and to the even more insidious idea inculcated by Coco Chanel that a tanned skin should be regarded as fashionable, and sexy, and a status symbol.

Some of the physical left-overs from a previous hairier existence remain with us (like the goose pimples), doing neither good nor harm, and in the long run will probably disappear. A more mysterious phenomenon is that of the sebaceous glands which cause anguish to so many adolescents by being the cause of greasy skin and acne.

The oiliness comes from a fatty substance called sebum. The sebaceous glands which secrete it are an appendage of the hair follicles, and the sebum seeps out onto the hair shaft and helps to keep the fur of mammals sleek and waterproof. A. M. Kligman, who made a special study of the subject, wrote:

> The original purpose was not so much to protect the skin as the hair. In man, however, save for a few

specialised regions, hair is a vestigial and rudimentary feature. With hair rendered obsolete, the sebaceous gland is literally out of work. It is a living fossil with a past but no future.

So it would be reasonable to expect that in man the glands would have dwindled to mere vestiges, just as over most of our bodies the hairs themselves have dwindled.

Instead of that, these oil glands have run riot. In our nearest relatives, the African apes, some sebaceous glands are found scattered over the body, but they are few and small. In man they are numerous and relatively enormous, especially on the face and scalp, sometimes extending to the neck and to regions of the upper part of the body.

An American enquiry into 'What good is human sebum?' reported that the short answer was 'No good'. It is not needed to keep the skin moist and supple: the softest human skin is that of a child, yet the sebaceous glands do not begin to operate until puberty. It used to be thought that the sebum probably helped to kill bacteria which landed on the skin, but that proved to be a fallacy. Yet from adolescence onwards our skin goes on producing sebum, not in response to any environmental stimulus (as is the case with sweat) but at a constant rate.

Globules of fat emerge from the sebaceous glands onto the surface of the skin, mingled with dead and decaying fragments of cells. The mixture is toxic to living tissue. At puberty the glands are growing so rapidly in the so-called 'acne areas' (face, chest and back of the torso) that the well of the hair follicle may become filled with a plug of sebum and cell débris. Or a cyst may be formed in which trapped bacilli produce irritant fatty acids and cause inflammation. The clinical symptoms — pimples, blackheads and inflamed nodules — are found most often on the face, but may extend to all the acne areas. On the scalp, where the follicles are less likely to become blocked, the same generous effusion of sebum provides a breeding ground for seborrheic dermatitis, otherwise known as dandruff.

This unmerited disaster may be visited on young people of either sex, but there are more victims among young males because males have larger sebaceous glands. The size of the sebaceous glands is influenced by the sex hormones (castrated males do not suffer from acne), and attempting to tinker with the balance of these hormones can have side-effects more dismaying than the pimples.

The effects may be partly mitigated by the use of antibiotics, but the only permanent cure is to grow older. With the passage of time the skin gradually adjusts to the situation, and the sebum is able to make its exit to the surface without battling its way out. After that the cosmetic results are less noticeable. Often they are confined to what the face powder advertisements used to denounce as 'shiny nose'.

It is cold comfort to a fifteen-year-old to know that the condition will not last, or to reflect that no other species in the animal kingdom has to put up with it. William Montagna, leading specialist on primate skin, commented: 'The human body appears to contain senseless appendages and even to make mistakes, but the sebaceous glands are too numerous and too active to be described as trivial.'

Other equally unexpected features of human skin remain to be discussed, but one thing should already be clear: a mammal cannot shed its coat of fur as a man can shed a suit of clothes and leave everything else unchanged. It is as revolutionary a change in life strategy as walking on two legs, and as far-reaching in its secondary effects. These complex changes would not have evolved, in our species alone, without some compelling reason.

6

Explaining Hairlessness

'In the water, fur provides poor insulation
and becomes atrophied.'
V. E. Sokolov: *Mammal Skin*

In 1989, before embarking on this chapter, I walked into
an Oxford bookstore to find out what the students in that
great seat of learning were being taught about this subject.
I was offered, and purchased, the latest editions of two
textbooks — both comprehensive, highly commended and
frequently updated to keep pace with the latest scientific
discoveries. One was called *Introduction to Physical Anthropology* and the other *Physical Anthropology*.

One of them devoted three words to the topic. By comparison with the apes, it conceded, humans 'have less
hair'. The other volume was more reticent and avoided
any mention of the subject whatsoever.

The writers are not falling down on the job. Their sole
object is to help students to pass examinations. They
know that the question of why *Homo* is hairless will not
crop up in the examinations, because neither the people
who set the papers nor the people who mark them know
what the answer is.

While that seems a little bit like a conspiracy of silence,
it is at least more dignified than the strategy in use some
twenty years ago. At that time the question was regularly
countered by a flat denial. Experts explained that people

who imagined humans had lost their body hair were being misled by a kind of optical illusion, because in fact humans have as many hair follicles all over their bodies as chimpanzees do — at least as many, possibly more. The fact that the hairs happen to be shorter, in some cases so short that they do not emerge above the surface of the skin, was dismissed merely as a 'quantitative difference' not requiring explanation.

This kind of evasion was an insult to the meanest intelligence, yet it was regularly uttered by professors and parroted by undergraduates. Desmond Morris copied down verbatim the pronouncement of one famous authority during a television programme: 'The assertion that we are the least hairy of all the primates is, therefore, very far from being true; and the numerous quaint theories that have been put forward to account for the imagined loss of hair are, mercifully, not needed.'

On television there has only once been anything approaching a debate on the aquatic theory. It took place in Chicago in 1972, and in the course of it a professor of anthropology was asked why man had less hair than the apes. Unprepared for the question, he earnestly assured viewers that there was no mystery about it: the answer had been found — the matter was well understood — the solution was there, on record, somewhere in the literature — but unfortunately it had slipped his mind.

For the benefit of any who might find themselves in a similar predicament, this chapter lists the explanations that have appeared 'in the literature'. Human nakedness has been attributed at various times to (1) ticks, (2) sex, (3) hunting, (4) neoteny, (5) noonday foraging, (6) allometry and (7) water. It will be for the reader to decide which of them is least implausible.

1   The parasite argument was advanced by a man called Belt, who argued that a naked primate would be less liable to harbour ticks and other noxious parasites which in the

tropics may constitute a serious danger to health. Darwin discussed this idea and dismissed it on the grounds that many quadrupeds are faced with the same problem but none have resorted to such a drastic measure of gaining relief. As an anti-parasite device nakedness would not only have been drastic, it would also have been ineffective. The flea that feeds on man is an exclusive species, recognisably different from all others, indicating that it has been our faithful companion over millions of years, and our lack of fur has failed to deter it.

2 Sex was Darwin's theory. His second book on the evolutionary theme was entitled *The Descent of Man and Selection in Relation to Sex*. He had not abandoned his belief that hairlessness constituted 'an inconvenience and possibly an injury', but he had collected evidence that many species had evolved features which were in themselves inconvenient or injurious yet were retained because they were attractive to the opposite sex.

Among his numerous examples were a species of lyre bird with a long, dangling tail which must hamper it in moving through the trees, and various brightly coloured male fish and birds whose adornment would make them more conspicuous to predators. These characteristics were nevertheless passed on because in those species the sexual adornment enabled the most spectacular males to attract more females and therefore to leave more descendants than their drabber rivals.

However, human hairlessness fails to fit into the normal pattern of sexual selection in several important respects. Firstly, sexual adornments are usually found in one sex alone and not in both. Secondly, they are usually found in the male and not in the female. Darwin believed that *Homo* was an exception, and that 'Man, or rather primarily woman, became divested of hair for ornamental purposes', and that the female subsequently transmitted the sexual adornment of nudity '. . . almost equally to their offspring of both sexes'.

He was able to cite examples of creatures where the females are more ornamented, but these are species in which there is total transposition of the sexual roles. An example is a bird called the phalarope, in which the female, beside being more brightly coloured, is also larger and more aggressive, engages in courtship behaviour to attract the male and fights to defend the territory, while the male tends the nest and hatches the eggs. No one suggests that this kind of transposition was manifested by the ancestors of *Homo*.

Thirdly, adornments for purposes of sexual selection are usually seasonal, more especially when they are inconvenient or injurious. After mating the lyre bird's tail and the buck's antlers are shed; the colours of the bright feathers and the shining scales are dimmed. A feature that was inconvenient all the year round, and to both sexes, would surely be selected out.

Fourthly, the characters which are sexually esteemed are usually, like the peacock's tail, additions rather than deficiencies. They are often indicative or suggestive of good health, such as lustrous hair or plumage, whereas the first stages of hair loss would have seemed more indicative of ill-health (mange, or debility, or alopecia) rather than fitness to survive.

Fifthly, there is in most people a rooted aesthetic prejudice against hairlessness except in the case of our own species. If we admired hairlessness in other animals, it would be easier to believe that our hominid ancestors yearned to find such a characteristic in their mates, but we find it almost universally repellent. We prefer the squirrel's bushy tail to the rat's naked leathery one, and the swan's feathered neck to the vulture's skinny one, the fluffy chick to the naked nestling, the velvet-coated European mole to the hideous naked Somalian mole-rat.

For these and other reasons, the idea of an ancestral

ape seeking out the females with the baldest bodies has fallen out of favour, even with Darwin's illustrious name to commend it.

3 The hunting hypothesis held the field for some time and was believed to explain many things about human physiology. The problem about hairlessness was why one savannah primate needed to go naked while all other species in the same habitat retained their fur. It was explained on the grounds that vegetarian primates do not need to move very fast, but a carnivorous primate would get hot while chasing its prey, and losing its hair would enable it to cool down.

The weaknesses of this hypothesis are, firstly, that fur or hair is a protection against over-heating because it provides a barrier against the heat of the sun. Secondly, that in the scenario of the hunting male and the non-hunting female, it would be predicted that males would become more hairless than females, whereas the reverse is the case. Thirdly, that any protection against over-heating in the day would have been counterbalanced by increased danger of hypothermia in the night. Fourthly, that on a predator-hunted savannah, all primates and not merely carnivorous ones would need to be able to run fast if they wished to stay alive, yet no other primate has lost its hair.

4 Neoteny is the concept that humans are a juvenilised form of ape. They are said to be characterised by a general retardation of the pace of development, so that they mature more slowly than the other primates and live longer, and this involves retaining some characteristics of a juvenile or foetal ape into adult life. The examples most often quoted when comparing humans with apes are the flatter face, the rapid pre-natal rate of brain growth persisting for some time after birth, and the hairlessness characteristic of a foetal ape. The theory suggests that the prolonged period of rapid brain growth was advantageous

to the species, and some other characteristics were retained as part of the neotenic package. One of these was nakedness, described as 'the embryonic distribution of body hair'.

The foetalisation theory was first proposed by the Dutch anatomist Louis Bolk in the 1920s, and revived in the 1970s by Stephen Jay Gould of Havard. Many people were convinced that the question 'Why naked?' now had a simple and adequate answer: 'Because of neoteny'.

The weakness of the theory is that while some characteristics may be retained as part of a neotenic package, that only applies to characteristics which are either benign or neutral in their effect on fitness to survive. No one claims that all foetal characteristics are retained in a neotenic species. For example, a human foetus and a human baby both have very short bandy legs, but natural selection ensures that this feature is not retained in adult life.

In 1977 Stephen Jay Gould published a definitive account of neoteny. The book was called *Ontogeny and Phylogeny*, and was rightly hailed as a classic. In it the author stressed that he had '. . . no intention of falling into the Bolkian trap of all or nothing'. He was concerned to give a clear and scrupulous account of his ideas on the subject which would stand the test of time. It is perhaps significant that in that volume the matter of hair loss is not discussed. The only fleeting reference to it is in a list quoted from a book by Louis Bolk. In summarising his own views, Gould stated his belief that neoteny does not inevitably result in the retention of juvenile characteristics, '. . . but it certainly provides a mechanism for such a result, if this result be of selective value.'

The last seven words are crucial. If nakedness increased the hominid's chances of survival in its particular habitat, neoteny would provide a mechanism by which it could be acquired. If it *decreased* the chances of survival in that habitat, then natural selection would countermand it and we would have remained hairy. Neoteny in no way absolves us of the need to identify the type of environment in which hairlessness is at a premium.

5   A more recent theory is that of the 'noonday ape', already discussed in connection with bipedalism. It hinges on the proposition, so far unproved, that an ape's brain may be more susceptible to over-heating than the brains of other primates like the baboon, and that *Homo*'s ancestors were therefore forced to adopt unique strategies to cope with the heat of the savannah. Its weakness lies in the idea that such an animal would opt to specialise in foraging during the heat of noon, a niche for which it would seem spectacularly ill suited.

The other speculative element is the widespread assumption that hairless means cooler. In its earlier and simpler form this was held to be self-evident: if a man feels cooler on taking his coat off, the same must apply to an animal. This had been disproved, so it is now replaced by a subtler proposition which states, (a) that some animals use sweating as a way of keeping cool, (b) that *Homo* is a prime example of a species making use of this strategy, and (c) that it is self-evident that the efficiency of sweating must be enhanced by hairlessness.

The weak link here is (c). It sounds so much like pure common sense that it is seldom questioned. One famous fossil-hunter on a radio programme was asked by a listener about human hairlessness, and he explained that when the hominids began to sweat, they were forced to become naked. He reminded listeners of the discomfort they feel when sweating while wearing a woolly cardigan, and told them that if the perspiring hominids had retained their fur it 'would have caused their skin to go mouldy'.

Human sweating is a phenomenon requiring a separate chapter, but one quotation may help to correct the mouldiness hypothesis. In 1974 a major survey of mammalian skin was published by V. E. Sokolov in the USSR, and an English edition by the University of California appeared in 1982. It described in detail the macro- and micro-structure of the skin of around 500 mammal species from all the orders and most of the genera, and followed that with a section on comparative morphology, and another on the way in which mammal skin is adapted for life in different environments. On page 578 Sokolov points out: 'It has been found that moisture evaporates twice as fast from fur as from a smooth surface, proving that fur does not prevent the evaporation of sweat.'

Many species combine sweat-cooling with a hairy pelt. None of them suffers from mouldy skin.

6 A new approach suggested in 1981 was based on the allometric analysis of hair density in primates. Allometry is the phenomenon by which, when species become larger in the course of evolution, not all organs of their bodies increase in the same ratio as their overall mass.

Allometric analysis shows that a monkey which is twice the size of another monkey does not have twice as many hairs to cover the increased surface of its body. In small monkeys like marmosets the hairs grow very close together, but increasingly massive primates have systematically fewer hairs per unit of body surface.

The allometric theory proposes that since the first hominids were descended from apes and apes are larger than monkeys, Lucy's ancestors would have had sparser hair and been less well protected against the direct heat of the sun than small savannah species like baboons and patas monkeys. Therefore, it is argued, they had to evolve an alternative system of keeping cool — that is, sweating — and that system ultimately became so effective that they were able to dispense with body hair almost entirely.

The argument is ingenious, but it ignores two facts. Firstly, a covering of hair is a multi-functional mammalian asset; it would not become redundant just because one function (heat regulation) was being otherwise catered for. And more to the point, decreased density of hairs to the square inch does not necessarily involve decreased coverage. A gorilla is larger than a man and its hairs are farther apart, but it is certainly not more naked. The mountain gorilla in particular has a deep and luxurious fur coat. If a savannah ape had needed a better shield against the sun's rays, its body hairs would have grown longer instead of shorter.

7   It was Sherlock Holmes's favourite axiom that when you have eliminated the impossible, whatever remains, however improbable, must be the truth. It is not to be expected that adherents of any of the preceding hypotheses will agree that their pet theories can be 'eliminated' from the enquiry, but the fact that there are so many of them proves at least that the question is still wide open.

The basic question is Desmond Morris's: 'In what environment is nakedness at a premium?' There would seem to be a simple and logical answer: it has to be the environment in which nakedness is known most frequently to have evolved. That environment is water.

Charles Darwin was saying the same thing when he wrote: 'Whales and porpoises, dugongs and the hippopotamus are naked, and this may be advantageous to them for gliding through the water; nor would it be injurious to them from the loss of warmth, as the species which inhabit the colder regions are protected by a thick layer of blubber.' The possibility that this could have any relevance to the human condition was not raised by him.

When Sokolov compiled his *magnum opus* on mammalian skin, *Homo sapiens* was one mammal not included among his hundreds of species, and when he wrote about adaptations to an aquatic environment he never counted how many of them could have applied to that illustrious absentee.

Fur depends for its efficiency as an insulating agency on its ability to trap a layer of air next to the skin. In land mammals, when the fur becomes wet, that ability is normally impaired and its value as an insulator is lost. Small aquatic mammals like the water shrew depend on especially oily fur, so waterproof that in effect it traps a bubble of air around the body and its insulative value is retained. Fur seals and sea otters have specialised pelts — an inner coat very dense and soft, capable of trapping tiny bubbles of air, and an outer layer of longer, glossy guard hair.

Larger and totally aquatic mammals, like whales, manatees and dolphins, have shed all their fur and replaced it by the fat layer more appropriate to an aquatic existence. Large and partially aquatic animals like seals and sea-lions which go ashore to breed, often in very cold latitudes, have of necessity kept the fur, which is the best insulator in air, and supplemented it with the layer of blubber which is the best insulator in water. But even among seals — especially those which have been aquatic for the longest time — the balance of benefit seems to tilt in the end towards hairlessness.

The largest species, the elephant seals (bulls measure 17–19 feet and weigh up to 8,500 pounds) spend about forty days out of every year completing their annual moult. Their skin is thick, and the process of shedding the dead cells which continuously flake off from it is hampered by the fur. The hair comes off in large strips and patches still attached to the outer skin, giving them an exceedingly moth-eaten appearance. R. M. Martin described them as looking at these times '. . . as if they were coming apart at the seams'.

For the second largest seal, the walrus, the attempt to hang on to a mammalian fur coat was obviously not worth the candle. It is completely hairless except for a rather splendid moustache.

Because the largest aquatic mammals are the most naked, it is sometimes urged that our own ancestors would have been below the body size at which hair loss becomes the best strategy in water. But there are several examples which counter that argument. One or two species of river dolphins are comparable in size to humans. So is the marsh-dwelling tapir — the nearest thing we have to an aquatic horse. It is an excellent swimmer and diver, has a proboscis like an incipient elephant's trunk, and its hair is very sparse. The most aquatic of the pig family, the babirusa — also smaller than a man — has no hair at all.

Only two kinds of environment are known to be conducive to nakedness in mammals — a totally subterranean one, like that of the naked Somalian mole-rat, and an aquatic one. No one suggests that our ancestors ever lived in a hundred per cent subterranean habitat.

There is one major barrier to acceptance of the idea that they may have lived at one time in an aquatic environment. The barrier is psychological rather than logical. It consists of the fact that the concept is new and bizarre and overturns at a stroke too many of the preconceptions we have grown accustomed to living with. In attempting to overcome that kind of resistance, reasoning alone in seldom effective in the short run. But the passage of time can work wonders.

7

Keeping Cool

'Sweating is an enigma that amounts to a major
biological blunder: it depletes the body not only
of water but also of sodium and other essential
electrolytes that are carried off with the water.'
William Montagna

In the above quotation Montagna was referring specifically to human sweating, which is in some respects unique. Some commentators speak of it as if it were an inexplicable burden we have to bear. Others praise it as an evolutionary innovation so brilliant that it was well worth losing our fur in order to improve its efficiency still further.

Sweating appeared at a relatively late stage of mammalian evolution. The first mammals — which appeared during the age of the dinosaurs — were small, furtive creatures about the size of shrews, and animals as small as that do not need to sweat. If the temperature gets uncomfortably high, they pant, and the moisture evaporating from their lungs and the tongue and the lining of the mouth provides an adequate cooling system. A large number of present-day species, including small arboreal primates like the loris and the potto, do not sweat. Some animals in hot countries supplement sweating by spreading saliva over themselves to moisten their skin, and this cools the body surface as it evaporates.

Homo sapiens appears to be the only land mammal which

does not pant on getting heated. Gorillas and chimpanzees do it — not as spectacularly as a dog, with noisy breathing and lolling tongue, but their rate of respiration certainly increases. People pant during exercise in order to inhale more oxygen, but they do not pant to reduce their temperature when lying down in the sun. We resort to it only in situations where sweating would not work — for instance, when sitting in a hot bath.

Later, as the mammals increased in numbers and diversified, some needed a special cooling system to supplement panting. For example, grazing animals have to spend long hours feeding in the open with no trees to shade them, so horses, sheep, cattle and donkeys all acquired the ability to sweat.

Some water can seep out through an animal's skin by diffusion, even when there are no sweat glands. But the amounts involved are very small, and inadequate for controlling body temperature. Water can pass out much more readily through tiny openings in the skin called pores. When the hooved animals began to sweat, they did not need to evolve special custom-made pores for the purpose: instead, they made use of the pores which had evolved to fulfil different functions. The history of evolution is full of such examples of physiological make-do-and-mend.

These openings already served three purposes. They were the holes through which the body hair emerged. They were also used as channels for the secretion of sebum for lubricating the hairs: sebaceous glands are found in association with the hair follicles. Thirdly, they were used as channels for the secretion of the apocrine glands, tiny structures situated near the base of the hair follicles. It is generally agreed that their original purpose was scent-signalling. They were designed to exude an oily or waxy substance which in various species and various bodily sites is often odorous, sometimes pigmented, and may contain proteins as well as oily substances.

Apocrines are highly adaptable. It is thought that even the milk glands of mammals are simply highly modified apocrine glands. In virtually all sweating mammals, with the exception of man, the so-called sweat glands are modified apocrines. The oily or waxy component has been diluted to a thin, watery emulsion, the consistency of skimmed milk. The glands are found all over the body in association with the hair follicles, and they respond to a rise in temperature by exuding this fluid onto the surface of the skin.

No one has described this kind of sweating as a biological blunder. The loss of fluid is not excessive, and is finely attuned to the animal's needs. In the case of the camel, for instance, only a thin film of moisture is exuded, to conserve water. In cattle, which use both panting and sweating to control their temperature, the evaporative heat loss through the skin is six times as great as the loss by panting, but at first this was not recognised because the sweat is not in excess of requirements and is not seen streaming down as it does in humans. Horses do sweat profusely when, for example, they are ridden at high speed, but that may be largely an emotional reaction. When at rest in a field even at high temperatures they do not stream with sweat.

In apocrine sweating there is efficient control not only of the amount of fluid loss, but also the loss of salt (sodium chloride). In many animals living on continental grasslands where sodium is in short supply in the environment, any salt in the sweat is largely reabsorbed. In sheep sweat, the sodium component can be replaced by potassium; in areas where sodium is particularly scarce the ratio of potassium to sodium may be as high as twenty to one.

On the supposition that man's ancestors moved out from the trees to open ground and needed sweat-cooling, they might be expected to have followed the example of the wild ass and the camel in adapting their apocrine glands for that purpose. They too would have needed a system not too profligate of water and salt — both scarce resources on the savannah.

But our ancestors did not make use of this thrifty and efficient method — and for a surprising reason. Humans cannot use their apocrine glands for sweat-cooling because, except for a few restricted areas of the body, they have lost them. Apocrines are found all over the body in most primates, up to and including the gorilla and the chimpanzee. Almost all over the human body they are missing. They begin to develop in the human embryo and are present all over the body of a foetus in the fifth month of gestation — but then they disappear.

The few remaining apocrines we possess are clustered in special areas and have never been adapted for temperature control. Their secretions are not diluted enough to be of any use for sweat-cooling. They are found in the armpits, the pubic area, the navel, the ears and the nipples, and they appear to have retained the original purpose of scent-production. For the young of most mammals, and perhaps too for human babies, the scent of the nipple is an aid in locating it.

Many primate species — and we are among them — have acquired specially evolved scent organs for communication, either to warn off rivals or to attract sexual partners. For example, a woolly monkey has scent glands on its chest, which it rubs against branches as a trade mark, laying claim to its territory. Lemurs have a cluster of apocrines forming an arm-pad near the inside of the elbow. They wipe off the scent with the opposite wrist and rub it into their long tails, which they then wave in the air to send the scent message far and wide to attract possible mates.

We (and the gorillas and chimpanzees) have a pad of apocrines structurally resembling the lemur's, but known to scientists as the 'axillary organ' (meaning 'situated in the armpit'). The apocrine glands there secrete a thick, greyish substance onto the surface of the skin; they are accompanied by other glands which supply a fluid to dilute the secretion, and pubic hair which helps to diffuse the aroma into the surrounding air.

These glands only become active at puberty, and their rate of secretion increases rapidly in response to emotional stimulation such as fear or sexual excitement. By analogy with other primates possessing such pads, it would seem that a primary function of the axillary organ was, originally, to act as an erotic attractant.

Something has clearly gone wrong. The resulting odour is more likely to keep a potential mate at arm's length than to lure one closer. This reaction is not the aesthetic affectation of a hyper-civilised population hooked on hygiene, and the proof of that is that it has gone wrong for the apes also.

Special scent glands are normally accompanied, as in the lemur and woolly monkey, by special behaviour patterns exploiting their potential. Some deer, for example, have scent glands just below their eyes and they perform the rather risky manoeuvre of depositing the scent onto twigs and thorns at a level where other deer will detect it.

Apes display no such behaviour in connexion with the axillary organ, nor is the armpit among the various parts of one another which they take delight in sniffing. It seems that at some point in their evolution they must have forgotten the axillary organ was there, or else lost confidence in its power to charm. Among our many inconvenient or embarrassing physical malfunctions this is clearly one that descended on us prior to the ape/man split.

When the axillary secretion is produced inside the gland it is virtually odourless. If our sense of smell were more acute we might even find it subtle and irresistible. But we are never able to encounter it in its pristine state, because the site has been invaded by a population of bacteria, which has been with us even longer than our fleas. The bacteria have settled permanently into the warm, moist, nutritious habitat of our armpits, feeding on its secretions and causing them to decompose instantly and emit noisome odours.

These local clusters of apocrine glands at armpits, pubic area, and so on, are all that remain in *Homo* of apocrines that in our ancestors appeared all over the body; all the others disappear before we are born. There has been remarkably little speculation as to why they disappeared. It is unlikely to be because we lost our hair. Even in areas of skin where the hairs are invisible, the hair follicles remain, and the sebaceous glands still make use of the follicles for channelling their secretions out to the surface of the skin. There is no physiological reason why the apocrine glands could not have continued to do the same thing.

When an organ atrophies in this way it is often because it has ceased to perform a useful function. If the original function, as is generally believed, was scent production, then the glands would not work so efficiently for an aquatic mammal because the pheromones — the scent-bearing secretions — would tend to be washed away in the water. In Egypt the water buffalo is a recent import, descended from an ancient swamp-dwelling species; it has only one-tenth as many apocrine glands as domestic cattle although they both belong to the same family, the *Bovidae*. It is often suggested that many aquatic species have dispensed with their sweat glands because in water 'they have no use for them', but there is great reluctance to believe that the same thing may have applied to *Homo*.

When the need arose for sweat-cooling in the hominids, it could not be effected by adapting the apocrine glands for that purpose because they were no longer there. Humans have adapted for this function another kind of skin gland — eccrine glands — which seem to have evolved originally to prevent an animal's feet from slipping.

Eccrine glands in all non-primate species such as wolves, lions, bears, mice, badgers, cats and dogs, are confined to the pads of their feet. They also differ from apocrines in other ways: they are never attached to hair follicles, but open directly onto the surface of the skin; they are active from birth and not only from puberty; they emit a clear, colourless fluid which is virtually odourless, with no oily component; its main constituent other than water is salt.

In arboreal primates eccrines are found on the soles of the feet and the palms of the hands where they perform the same function of protecting against slipping. Clearly they are more important to primates than to ground dwellers. If a dog's foot slipped when it was climbing up a rocky slope, the worst consequence would be an undignified slither down to the bottom, but if a monkey's hand slipped when it was hanging onto a branch, it could crash to the ground.

The hands and feet of monkeys and apes and humans are doubly guarded against slipping. They have ridges, like the whorls on our fingers, which improve the grip just as the treads on a tyre do. And they have the eccrine glands which dampen the skin, and damp skin provides a better purchase. We demonstrate our awareness of this when, for example, we moisten a finger-tip when turning over a page, or counting bank notes, or when a gambler licks his thumb before dealing cards.

Some primates therefore developed eccrines in additional places. Some monkeys in South America (though not in Africa or Asia) use their prehensile tails as an additional anchor for hanging on to the branches, so that the underside of their tails is naked, and the skin is ridged and well supplied with eccrine glands. Chimpanzees and gorillas walk on their knuckles and their knuckle pads bear 'finger-prints' and eccrine glands.

None of these examples has anything to do with sweat-cooling. A monkey's palm sweats, not in response to a rise in temperature, but in response to a consciousness of danger. When the monkey takes a decision to launch itself into space to leap from one branch to another, the brain sends out a signal which quickens its heartbeat, sharpens its perceptions, and at the same time dampens the palms to ensure a good grip on the branch it is aiming for.

Our own palms sweat in exactly the same way. They do not respond to changes in temperature. Instead, the wetness breaks out when we are tense or apprehensive — standing in the wings with stagefright, being introduced to someone we are in awe of, or contemplating a crucial shot in a snooker final so that the hands have to be dried on a cloth before cueing. Sometimes the moistening is overdone. In one person in a thousand sweating palms are troublesome enough to require medical treatment — but normally it is only a minor inconvenience.

Unless, perhaps, you are being interrogated. Even a slight dampness of the palms increases the conductivity of your skin, and it may be registered and recorded by a polygraph — the so-called 'lie detector'. If you think: 'I am about to tell a lie and perhaps they will find me out,' the machine will detect your alarm at the preconceived risk. But the machine does not detect lies — it only detects anxiety. You may feel a stab of apprehension before embarking on a perfectly true story if you know it sounds too improbable to be believed. On the other hand, if you are a hard-boiled type with a hundred per cent certainty of never being found out, you may lie your head off and still pass the test.

In the course of primate evolution some eccrine glands began to appear scattered at random over the body surface. In small species they are comparatively few and apparently 'ectopic' — simply occurring by chance in the 'wrong' place. In larger monkeys there are more of them. By the time we get to the African apes they are as common as apocrine glands, or slightly more so, in a proportion of about 52 to 48 per cent. In humans the eccrines have made a quantum leap, to something nearer 99 to 1 per cent.

As far as anyone can discover, the apes do not use them for anything at all. Humans use them for sweat-cooling.

Compared to our naked skin, this is not a phenomenon that arouses popular speculation, because it is invisible to the naked eye. But it has always been a source of amazement and bafflement to specialists who spend their lives studying mammal skin — from P. Schiefferdecker who said in the '20s: '*Homo sapiens* may accurately be described as the eccrine gland mammal', to V. E. Sokolov, who declared in the '80s: 'Eccrine sweating is a human characteristic as unique as speech or bipedalism.'

It is the specialists, on the whole, who are least likely to proclaim eccrine sweating as a great step forward. Everything points to the fact that it is a fairly recent development, certainly arising later than the ape/man split. Deficiencies in the system may thus be due to the fact that it is still in the running-in period. As William Montagna has observed: 'The several million (eccrine) glands on the human body act principally as heat regulators, but the function is perhaps too recent to be totally effective.'

It is inefficient in four ways: (1) it is slow to start; (2) it is wasteful of water; (3) it is wasteful of salt; (4) it is slow to respond to danger signals when the body's resources of water and salt are getting low.

The slow start is the cause of the phenomenon known as 'sunstroke'. Unlike the palm sweat of embarrassment which can respond within seconds to the perception of a tricky situation, the heat regulating eccrines take around twenty minutes or even longer before they respond to a rise in temperature. Within that period the body's internal temperature, including the brain temperature, may have climbed high enough to impair brain function and cause a sudden collapse. A typical example would be that of a cricketer walking out of a cool pavilion on a very hot day, or soldiers marching onto a parade ground to stand to

attention in blazing sunshine. However strict the discipline, there is always the possibility that two or three of
them will keel over in a dead faint.

When the sweat finally breaks out, the body temperature comes down very rapidly. It can descend farther and
faster in a short time than in any other animal, and this
is sometimes cited as proof of the superiority of our own
peculiar system. It could equally well be a sign that the
temperature had climbed to heights it should never have
been allowed to reach in the first place.

Human sweating is also praised because it breaks all
records in respect of its profusion, as if there could never
be too much of a good thing. This is a fallacy. All that is
needed to cool the skin is a thin film of moisture such as
that produced in the sweat glands of the camel. A more
copious supply does not do the job any better. The truth
of this is brought home to anyone who has been called
out of a bath to answer the telephone in a draughty hall.
A perfunctory towelling of the limbs does little or nothing
to reduce the chill factor, which only wears off once the
skin is completely dry — proof that a very thin layer of
moisture cools the skin quite as effectively as a more
copious one.

The sight of an athlete, or a traveller in a hot climate,
with sweat streaming down the skin and dripping off
onto the ground is not the sign of an efficient system, but a
wasteful one. Sweat surplus to requirement serves no useful purpose. Among New Yorkers the commonest gripe
about their summer climate is: 'It's not the heat — it's
the humidity.' When the surrounding air is too moisture-
laden to allow evaporation, sweating affords no relief.
The continuing pointless flow of perspiration only adds
to their discomfort and is no better than a nuisance.

In some circumstances it is much worse than a nuisance.
Desert fighting in the Second World War stimulated
research into the question, and it was found that a soldier
engaged in strenuous activity in a hot climate may lose

up to ten or fifteen litres of water per day through his skin, and he cannot survive for long unless all that is replaced. For anyone cut off or left behind with limited supplies of water, profuse eccrine sweating would greatly accelerate the onset of dehydration and death.

It seems extremely unlikely that eccrine sweat-cooling would have evolved in an ape living on the savannah, where water is precious. Having to return frequently to one of the widely scattered water holes would have limited the range of its foraging and increased its vulnerability to the predators that frequent such places. And, unlike the camel which can drink 30 per cent of its body weight without stopping, humans are unable to drink enough to supply their needs for several days. For an animal of our size we are near the bottom of the league in the amount we can drink at a time. One scientist, R. W. Newman, summed up our predicament in a memorable sentence: 'Man suffers from a unique trio of conditions: hypotrichosis corpus, hyperhydrosis and polydipsia' — which is to say we have a subnormal supply of body hair, a supernormal output of sweat, and are under the necessity of drinking much and often.

The other precious commodity which is squandered during profuse sweating is salt. In the interior of continents, salt in the soil and vegetation is often in poor supply. Over millions of years rainfall and erosion and rivers have been transferring more and more of the world's resources away from the continental masses and into the sea. Some is returned in the rain, but the salt in rain water decreases rapidly with distance from the ocean. Over much of Africa's grassland the kind of sweating which causes salt to crystallise out onto the surface of the skin would be a heavy liability to a hominid.

Salt deficiency causes debility, impairment of functions and heat cramps. It was J. B. S. Haldane in the 1920s who discovered that the violent attacks of abdominal cramp which affected miners, stokers and steelworkers were due to prolonged sweating. He advised ship's stokers in hot countries to add ten per cent sea water to the water they drank, and that cured the problem. At maximum sweating capacity the body can lose its entire sodium pool in just three hours, ultimately leading to death.

It is strange, therefore, that man does not diminish sweating in the heat in response to signals of dehydration or salt deficiency until these have reached dangerous levels. This failure does not merely affect the athlete at the end of a marathon, who may be in a state of mental confusion and physical collapse.

Mortality tables show that even in temperate countries like Britain a heat-wave is invariably followed a day or two later by a peak of deaths from thrombosis. During one heat-wave in Greater London the daily thrombosis death rate doubled, and the same phenomenon occurs every year in the United States. It happens because even when water is freely available, the fluid lost by sweating is not replaced. The blood volume goes down, the platelet count goes up and the plasma cholesterol rises. These are precisely the conditions liable to cause blood clotting, with the consequent danger of stroke or heart attacks in elderly people or others with high blood pressure. Deaths from this cause are much commoner than the more dramatic collapse from heat stroke.

In view of these facts it is hard to understand the savannah theorists' assertion that the hominid's unique system of thermoregulation achieved new 'high levels of efficiency'. In the severe heat-wave in 1987 in Greece it was not the goats on the hillsides that fell victim to it. With their protective coats and low-output apocrine sweat glands, they were survivors. But thirteen hundred of the people died.

8

Sweat and Tears

'The more you cry, the less you pee.'
Folk saying

A sweat system prodigal of water and salt and leisurely in its response to rising temperature suggests an environment with an ample supply of water, an ample supply of salt and not much threat of over-heating — that is, somewhere wet and cool and briny. But none of the research on the subject addressed itself to the question of 'Where would this development have been at a premium?'

For a long time it was common practice to describe the apocrines used by other animals as the 'pseudo sweat glands', and eccrines as the 'true sweat glands'. The implication seemed to be that in this respect all other animals are out of step except *Homo*.

Since our eccrines functioned as temperature regulators, it was baffling that the same glands in other animals refused to perform the same service. Many experiments were conducted with animals to try to induce eccrine sweating by raising the ambient temperature. Favourite subjects were cats since they were readily available, and chimpanzees because of their close relationship to ourselves.

The results were negative. Neither the cat's paws nor the chimpanzee's body could be induced to sweat within the range of temperatures they would ever encounter in the wild. At higher temperatures a few drops of moisture were observed around the eyes, lips and scrotum of a chimpanzee. But that is not necessarily evidence of eccrine activity. A shot of adrenalin produced similar results at the same sites in a horse, and a horse has no eccrine glands at all.

Since the 1970s emphasis has moved to studying cooling strategies in savannah primates. All use panting as a means of relief and some — for example, baboons and patas monkeys — supplement this by sweating.

The patas, reputedly the fastest-running primate, under laboratory conditions in high temperatures with drinking water freely available, has been observed to exude sweat at a rate of up to 50 per cent of the human maximum per square centimetre of skin. In the wild they are better able to afford sweat cooling than other savannah species because they feed largely on fruit instead of almost exclusively on grass and roots.

The question is whether this demolishes the long-held belief that human thermoregulation is as unique to our species 'as speech or bipedalism'. Like the apes, savannah primates do have some eccrine glands on body and limbs as well as on hands and feet. Hence there is a disposition to assume that their sweat-cooling, like ours, must be eccrine.

But this remains an area of contention. The difficulty is that in these animals the two types of glands are closely interspersed. This makes it very difficult to design an experiment which both measures the sweat and ascertains which type of gland it comes out of. A series of experiments with the patas monkey confined itself to measurement alone. In one experiment with a rhesus monkey the sweat was described as eccrine; in another experiment with baboons the verdict was apocrine.

It is perhaps unnecessary to point out that in these monkeys the sweating is in no case 'facilitated' by the shedding of body hair. The patas, in fact, has a thicker coat than some arboreal members of the *Cebidae*.

The strength of the aquatic case lies in the fact that those features of the human skin that are *undeniably* unique among primates — the nakedness, the underlying fat, greater elasticity, dearth of apocrines, proliferation of sebaceous glands — can all be paralleled in aquatic species, and none in grassland species.

A study of sweat and sebaceous glands in seals by J. K. Ling reports that '. . . with progressive hair loss, the need for waterproofing the skin is met by enlarged lipid-secreting sebaceous glands.' Waterproofing seems to be the only function of sebum. In the apes, sebaceous glands are consistently found in those parts of the body which are always, or frequently, wet and at the same time in contact with the air — around the mouth, along the edges of eyelids, around the anus. In a wading ape finding some of its food on the sea bed, waterproofing would be most essential in those parts of the skin which would be most frequently wetted and then exposed again to the sun and wind, namely, the head and face and upper parts of the chest and back. These are precisely the locations where our sebaceous glands are found.

There is even one clear aquatic instance of eccrine sweat-cooling. When the fur seal goes ashore to breed it suffers from the heat because of its double insulation of fat and fur. It is believed to be descended from a possibly dog-like land-dwelling ancestor with foot-pads of the standard mammalian type: the flippers are covered with eccrine glands which sweat profusely as it waves them in the air to cool itself.

If, then, we provisionally adopt the aquatic hypothesis, we are confronted with some entirely new questions. Several dermatologists have voiced the conviction that our sweat glands must have originally evolved for some other purpose, without putting forward any suggestion about what that purpose could have been.

What purpose could they have served in the sea? All they are structurally designed to do is to emit a saline solution — water with salt dissolved in it. There are trace elements of urea, and of ammonia (probably due to bacterial decomposition), but the main constituent — 90 per cent of everything except the water — is salt.

Birds and mammals which get their food from the sea sometimes take in more salt than their bodies need, and too much is just as harmful as too little. Many species have had to evolve special ways of getting rid of the surplus. The most familiar example of this is found in the salt glands of seagulls.

From time to time dermatologists have referred to the possibility that the non-volar eccrine glands (those not found on the palms and soles) originally served the purpose of excreting salt. The names first given to the skin glands by Schiefferdecker reflect his conviction that the eccrine glands were excretory in function. 'Apocrine' indicates that the secretions, such as oil or proteins, were made by or out of the gland itself; 'eccrine' means that the sweat was being passed out through the gland without having been manufactured in it.

(In recent papers the terms apocrine and eccrine are often replaced by 'epitrichial' and 'atrichial' respectively, meaning 'attached to the hair' and 'not attached to the hair', but as a general rule changes in established terminology are better avoided. For example, it would be easy to invent an apter name for the chimpanzee than *Pan troglodytes* — 'cave-dwelling satyr' — but the only result would be confusion.)

Whenever anyone suggests that the eccrine glands evolved to excrete salt, the idea has been discounted for one simple reason. Human sweat is hypotonic — that is, it is a saline solution more dilute than the blood. To serve any useful purpose in eliminating salt it would have to be, like urine, saltier than the blood.

It is quite possible, however, that at an earlier stage

in our evolutionary history the excreted fluid was more concentrated than it is now. Even today, when sweating is prolonged, it becomes saltier. In scientific terminology the sweat gland becomes 'fatigued', or 'progressively suffers a loss in its capacity to produce a hypotonic fluid'.

This capacity to keep the fluid dilute may have been a comparatively late evolutionary development. The energy for keeping the fluid hypotonic is obtained by converting glycogen in the glands into lactic acid. But this only happens in the more recently evolved eccrines on our body and limbs. In the more primitive eccrine glands of our hands and feet this conversion does not take place.

All that is highly speculative. But there is one piece of persuasive evidence that our ancestors did at some point in our evolution encounter a salt crisis and needed some extra mechanism, in addition to the action of the kidneys, for eliminating salt.

Marine birds which catch and swallow salt-water fish inevitably swallow some sea water in the process. Some sea food such as squid is intrinsically salty. Drops of clear fluid can sometimes be seen dripping from the birds' beaks. On investigation this has proved to be a concentrated salt solution coming from specially evolved nasal salt glands, and the largest salt glands belong to those species which spend most time at sea.

Marine reptiles, and some marine mammals, encountering the same problem, deal with it not by nasal dripping but by weeping salt tears, for example, marine iguanas and turtles, marine crocodiles, sea snakes, seals and sea otters.

' "I weep for you," the Walrus said, "I deeply sympathise," holding his pocket handkerchief before his streaming eyes.' Lewis Carroll, author of *Alice's Adventures in Wonderland*, wrote fantasy, but he got his biology right. The only weeping creatures in his books are the Walrus, the Mock Turtle — and Alice. (His crocodile was not a marine one but lived in the Nile, so it remained dry eyed and 'gently smiling'.)

Man is the only weeping primate. With our copious sweat and our copious tears, we are by far the leakiest of all the apes. Because the phenomenon is so hard to explain it is often dismissed (as nakedness used to be) as non-existent or 'merely quantitative', a slightly increased activity of the lacrimal glands which nearly all land mammals possess.

That is not true. Emotional (or 'psychic') tears are quite unlike the tears which coat the eyes with a film of moisture and flow more freely in response to irritating vapour or a foreign body in the eye. Psychic tears respond to different stimuli and are controlled by different nerves. If the root of the trigeminal nerve which leads from the brain to the eye is severed, ordinary reflex weeping in response to irritants is prevented, but tears of grief or joy will still flow. The same phenomenon is observed if cocaine is applied to the surface of the eye.

The connection between weeping to excrete salt and weeping from emotion is not easy to understand, but it is an ancient one. There are accounts of copious nasal dripping in seagulls in situations of aggressive confrontation, and weeping in sea otters when distressed or frustrated. The stimulating hormone prolactin, which appears to be involved in human weeping, is released in response to emotional stress.

Like human sweat, human tears are hypotonic, but may not always have been. Tear glands, too, can get 'fatigued' if weeping goes on long enough; the salt may become more concentrated and can sting the eyes. Shakespeare in *King Lear* described a depth of grief in which 'mine own tears did scald like molten lead'. It seems quite possible that eccrine sweating and psychic weeping evolved at the same time. When malfunctioning occurs, it is liable to affect both equally. In cases of cystic fibrosis, for example, one of the diagnostic signs is extra saltiness both of sweat and tears.

Not much is known about psychic weeping because it is very difficult to research. The phenomenon does not occur in any of the normal laboratory animals; *Homo sapiens* is the only available subject. Having obtained a willing volunteer, the scientist is still confronted with the problem of how to induce the weeping. Reflex tears are easily stimulated: onion vapour, which turns to sulphuric acid on contact with a moist eyeball, can be administered in controlled doses. But not all volunteers can weep emotional tears to order.

Darwin's account of weeping in *The Expression of the Emotions in Man and Animals* drew heavily on observations of young children. He had a large family and when the children were young and tearful he studied them closely in the short interval before the paternal instinct to comfort them triumphed over scientific curiosity. He was among the first to note and record that in the first few weeks of life a baby's vocal cries are unaccompanied by tears, although reflex tears are operative from birth.

The theory he finally advanced was that when a child has a screaming fit, this leads to engorgement of the blood vessels of the eyes (as does, for example, vomiting). That in turn leads to spasmodic contraction of the muscles around the eyes in order to protect the blood vessels against the danger of bursting, and he believed the resulting pressure could activate the lacrimal glands. He conceded that adults often shed tears unaccompanied by screaming or screwing up the eyes, but suggested that by then an association between heightened emotion and weeping has become automatic.

A leading present-day authority on the subject, William Frey, failed in his first attempts to collect psychic tears. Volunteers who claimed to be able to weep whenever they wished — including an actress who had repeatedly demonstrated this ability on stage — were inhibited by the clinical atmosphere of the laboratory.

The problem was only solved when Frey invited an audience of volunteers to watch 'tear-jerker' movies and collect their tears in test tubes. In ancient times similar vessels called lachrymals were specially made for the purpose, and Nero was said to have shed a copious supply while he watched Rome burning.

Some interesting discoveries have been made. Frey found that the chemical composition of the two kinds of tears is different, with a protein content at least 20 per cent higher in emotional tears than reflexive ones. He established that, contrary to the traditional belief, lacrimal glands are capable of excretion. For example, the element manganese — deleterious in high levels — is found in tears at a concentration 30 times higher than its concentration in blood. He believes that one modern function of tears may be to eliminate excess stress-related chemicals, and that that is why weeping can bring emotional relief. He also found that the tear-generating emotions aroused by sad movies were more containable when the viewers were in physical contact with each other.

As in many other instances, scientific investigation confirms what most people instinctively know. Confronted by children or friends in a state of distress, they feel the impulse to establish physical contact by embracing them, or else to utter the familiar piece of folk wisdom: 'Go on, have a good cry — you'll feel better.' Frey has shown that the instinct is sound. As far as I know, no one has tested the validity of the piece of folk wisdom at the head of this chapter. But it is known that the excretion of urine diminishes during copious sweating, so there could conceivably be an element of truth in the old adage.

There is one final indication that psychic weeping may have had a primitive connection with something undesirable being swallowed — for instance, too much sea water. That is the lump in the throat which is part of the weeping experience and often precedes or accompanies bursting into tears. It has been described as being like 'a football in the throat', and early physicians concurred with this description by naming it the 'globus hystericus'. In a survey of 331 people, females reported having a lump in the throat in 50 per cent of crying episodes, and males in 29 per cent. It is caused by a cricopharyngeal spasm — an involuntary muscular contraction which blocks off the entrance to the gullet, temporarily preventing anything from entering the stomach. No other explanation of this phenomenon has ever been attempted.

A salt crisis in our evolutionary history would go far to explain one other specifically human characteristic — the fact that we have no instinctive awareness of the state of the sodium balance in our bodies. When we are suffering from a deficiency of water we feel thirsty, and take active measures to find a source of water and set the balance right. If the thirst remains unslaked it intensifies until all other considerations become subordinate to the need to drink.

Most mammals respond just as urgently to the need for salt if they are deprived of it. Derek Denton, in his classic study *The Hunger for Salt*, describes his researches into salt appetite in non-human mammals such as sheep, rats and rabbits. In all these species there is a precise correlation between the amount of salt their bodies need and the amount they will take in.

Species living in habitats far from the sea go to great lengths to satisfy their salt hunger. Grazing animals will make long treks to locations where volcanic action, or the relic of an ancient sea bed, have left deposits of salt in the earth. At Mount Elgon in the African Rift Valley there are caves which harbour salt-rich soil. Herds of elephants journey to these caves and venture deep into the darkness, as many as ten at a time, to dig out and eat lumps of the earth.

Some areas are so naturally deficient that man himself

may be an important source of salt. Climbers in mountain areas have found that wild sheep will be lured to lick their clothes where they have been saturated with perspiration. In the wooden houses of parts of British Montana, men had to abandon their habit of urinating against their verandah posts because porcupines came out of the woods at night and were liable to chew right through the posts for the sake of the salt in the urine-soaked timber. Gorillas and chimpanzees are among the animals which have been observed to dig and eat earth for the sake of its mineral content.

On the other hand, when an animal has had enough salt it will take no more. In humans neither the compulsory search nor the abrupt cut-off point can be relied on. Their intake bears no relation to salt deficit or surplus. This is surely not a characteristic that would have been acquired on the savannah. As Denton points out, in savannah conditions '. . . selective pressures would favour retention and perhaps elaboration of salt mechanisms'.

J. B. S. Haldane was one of the first to suspect that this was a specifically human defect. In a paper based on his researches into heat cramps, he reported:

> We may, if we please, regard the symptoms due to shortage of sodium chloride as due to failure on the part of the nervous system, since miners, stokers, etc., can obtain as much salt as they 'like'.

The problem was solved by adding salt to the workers' food or drinking water. Much more recently a similar 'miracle cure' was found which saved millions of lives in the Third World: oral rehydration.

It is an extraordinary fact that what *The Lancet* described as '. . . probably the most important medical breakthrough of the century' was one of the simplest and cheapest. By this time it has saved millions of lives in the Third World. As the UNICEF report pointed out in 1984:

'These children do not die from exotic causes requiring sophisticated cures. Five million of them die annually in a stupor of dehydration caused by simple diarrhoea.' Salt eliminated in the faeces left their bodies short of sodium. Oral rehydration therapy – a simple cheap solution of sugar and salt — instantly halved the diarrhoea death rate in the villages where it was administered. No other animal needs a professor or a travelling nurse to make it aware that it is suffering from a salt deficiency, or to issue advice on what to do about it.

At the other extreme, people in the First World are liable to consume up to fifteen times as much salt as they need, and the same 'failure on the part of the nervous system' leaves them unaware that they are overdoing it. Many doctors have told us that statistics from many countries seem to suggest that a high intake of salt correlates with a high incidence of hypertension. There are tribes living in low-salt areas where blood pressure actually goes *down* in old age. Among white races, high salt intake has been identified as a factor in hypertension for some people, though not for everyone. But the ill-effects of over-consumption of salt are less dramatic and immediate than the results of deficiency, so their warnings to date have made little impact on our eating habits.

An ape living on sea food would have been comparatively well placed to adjust to a saline environment. As a primate it possessed eccrine glands in greater numbers than non-primate species, and they were already structured to exude salt water. All that was needed was an increase in the output of the glands and a further increase in their number. The instinct for responding to 'salt hunger' could well have been lost at this stage. Sodium deficiency would not have been a danger that needed guarding against.

As the land-locked sea continued to shrink and become saltier, the aquatic apes would have been driven to retreat south along the water courses of the Rift Valley. By the time they were ready to return to terrestrial life, apocrine sweating was no longer an option for them: they had lost their apocrine glands. The eccrines were drafted to fill the gap, needing only two modifications. They have been adapted to respond to heat instead of salinity and to keep the leakage of sodium within more acceptable limits.

9

Fat

'An exceptionally obese monkey resembles
a lean human being.'

Caroline Pond

In 1930 a young marine biologist called Alister Hardy
returned from an Antarctic expedition and read a book by
the eminent anatomist Professor Frederick Wood Jones.
In it he came across the following passage:

The peculiar relationship of the skin to the underlying fascia is a very real distinction, familiar enough
to anyone who has repeatedly skinned human subjects and any other member of the Primates. The
bed of subcutaneous fat adherent to the skin, so
conspicuous in Man, is possibly related to his apparent hair reduction; though it is difficult to see why,
if no other factor is involved, there should be such a
basal difference between Man and the Chimpanzee.

Hardy was forcibly struck by what he read because he
knew from first-hand experience that just such a layer of
fat, bonded to the skin, was a common characteristic of
most aquatic mammals.

These facts were, of course, familiar to biologists. But
it took a giant leap of Hardy's imagination to bridge the
gap and arrive at the daring conclusion that the ancestors

104

of *Homo sapiens* may have acquired this fat layer in the same way as the whale and the seal and the penguin and the dolphin and the hippopotamus — as an adaptation to an aquatic environment during some part of their evolutionary history. He felt that such an interlude would at the same time account for other unexplained features of the human anatomy. The fat layer was thus one of the earliest of the arguments advanced against the savannah hypothesis. It has never seemed very plausible that being fatter could be an advantage to hominids in a savannah habitat. Whether chasing prey or running away from predators, the extra weight would be bound to slow them down.

The fat layer, then, is a mystery, but a mystery made easier to ignore by the wide disparity between individual human beings. Some are fat and some are not fat. There is a tacit assumption that the leaner ones are the norm, and illustrate the way Nature always meant us to be. If some humans exceed the 'norm' it is assumed they are too greedy or too lazy or suffering from some metabolic disorder. Fatness is never acclaimed as a human attribute in the way the large brain is acclaimed. It is spoken of as if it were a pathological departure from the true human blueprint, as something to be treated and, if possible, 'cured'.

First, therefore, it is necessary to establish that subcutaneous fat in *Homo sapiens* is a specific evolutionary development and not simply a punishment for eating too much. One way of doing this is to consider the case of a human who cannot be accused of gluttony, that is, a new-born baby.

Around the thirtieth week of gestation, the growth pattern in a human foetus begins to change. The hitherto rapid growth of the bones begins to slow down, while resources are concentrated on the accumulation of fat. Some of the fat is deep inside, in standard mammalian sites such as that around the kidneys. But it also accumulates under the skin all over the body, in a way uncharacteristic of land mammals.

Between weeks 30 and 40 the amount shoots up from 30 grams to 430 grams. At full term, fat constitutes sixteen per cent of body weight in a human baby, as compared with three per cent in a new-born baboon. The more fat the baby has acquired by the time it is born, the greater its chances of survival. Moreover — like brain size — the fat deposit sustains its rapid growth rate for several months after the birth. This — more than brain size and just as much as hairlessness — is what distinguishes the appearance of a human baby from the infants of the other primates. A week-old human baby is naked and plump; a week-old gorilla or chimpanzee is hairy and, by comparison, cadaverously thin.

The laying down of this fat layer in the baby imposes an extra strain on a human's physical resources during late pregnancy and lactation which other primates do not have to sustain. The percentage of lipids (fat) in the mother's blood goes up by more than 50 per cent to facilitate the transfer of food resources to the growing foetus. To replace these resources she needs to increase her own food intake by fourteen per cent in the last stages of gestation, and by up to 24 per cent while breast-feeding a growing infant. If she does not do so, the reserves of fat (and calcium, and whatever else is in short supply) will be drawn from her own body. Nature is almost always on the side of posterity, so that even if there is a food shortage during pregnancy, it is unlikely to reduce the birth weight of the baby by more than ten per cent.

However, if the woman were seriously undernourished, the chance of her giving birth to a viable infant, or herself surviving the pregnancy, would be infinitesimally small. This contingency is guarded against by ensuring that if a woman's fat reserves are below a certain level she will not conceive. In a 16-year-old girl, fat tissue constitutes around 27 per cent of body weight; if it drops below 22 per cent, menstruation will not begin and, if it has begun, it will cease.

This does not only apply to women and girls who are anorexic, or have become ill and debilitated through severe malnutrition; it also applies to healthy, active females if they happen to be ballerinas or athletes on a regime of eliminating all 'surplus' fat. If and when their weight is allowed to go up, the periods normally recommence.

It seems clear, then, that a percentage of fat well in excess of the primate norm is not merely non-pathological in our species: it is essential to the survival of the human race.

Today in the developed countries few people need to worry about the minimum fat level which natural selection has laid down for our species. But much anguish is caused by its apparent failure to lay down a maximum level. If a horse, a wolf, a cheetah or a kangaroo is kept in captivity with little or no opportunity for exercise and given more food than it needs, it may put on some fat. But it will not treble or quadruple its body weight. A non-human primate kept in captivity may develop a rounder abdomen, but it will not develop bloated cheeks, fat buttocks, flabby arms, a large bust and fleshy thighs.

Humans are liable to develop these characteristics because one of the main depots in *Homo sapiens* is just under the skin. In most mammals the main depots are internal, so their expansion is limited in extent by the body wall or the rib cage; the result is also less visible. But our skin is so elastic that it imposes virtually no limit on the amount of fat that can accumulate there.

Another notable difference is the number of adipocytes we are endowed with. Adipocytes are the cells which contain the fat. When they are empty they are virtually flat, but each one is capable of swelling up, becoming spherical, and expanding to three times its original size without bursting. Hence a major factor governing how fat we can get is the number of adipocytes we possess.

A survey was carried out of 191 different mammal species. It was found that carnivorous mammals in general have more adipocytes in proportion to body mass then herbivores. But the most remarkable feature of the final report was summarised by Caroline Pond, senior lecturer in Biology at the Open University in Milton Keynes:

> However we compare them, *Homo* is clearly the odd one out. In proportion to body mass, we have at least 10 times as many adipocytes as expected from this comparison with wild and captive animals. Humans easily surpass such notorious fatties as badgers, bears, pigs and camels, and are rivalled only by hedgehogs and fin whales in their deviation from the general trend.

The hedgehog and the fin whale exemplify two kinds of mammal which need extra adipocytes, namely, the hibernators and the aquatics. No one would seriously argue that we acquire our ten-fold allowance of those cells because our ancestors went through a stage of being hibernators. Hibernators only get fat seasonally, just before retiring for their winter sleep, whereas fat is with us all the year round. And if we cut out the hibernators, the picture is clear. An excess number of adipocytes is among the characteristics which man shares with aquatic mammals.

There must be a reason for the existence of these cells — about twenty-five billion of them, ten times as many as in other land animals of our size, and ten times as many as we presently need. They must have been acquired because at some stage of our evolution they were useful or necessary. The unavoidable conclusion is that at some stage in the past our ancestors were considerably fatter than we are today.

Nowadays people who are 'overweight' are often made to feel a kind of guilt, as if they were degenerate heirs of an endless line of lithe and slender ancestors — as if they had somehow betrayed that inheritance by allowing themselves to go to seed.

They have not betrayed their inheritance. Their inheritance has betrayed them. They come into the world with a capacity to become obese not shared by other primates. If the excess adipocytes were not present in their bodies they could not be filled up with fat. If we had only one-tenth of that number, then once those receptacles were full any further excess calories would perforce be burned away by a rise in temperature, or excreted as fat in the faeces or as sugar in the urine, and not added to the deposits on the belly or buttocks.

Cheerleaders in slimming clubs are apt to prescribe for their clients a certain number of kilograms as representing the amount they 'ought' or 'were meant' to weigh. They urge weight watchers to aim at the target and become slim as Nature intended. But the sad fact is that 'Nature' is not on the side of the slimmers. Nature decrees that when they have lost the first half-stone, their metabolism slows down and calories are burned up at a slower rate, which makes it that much harder to lose the second half-stone. If they achieve their target they are entitled to take credit for having succeeded not with Nature's aid, but by dogged determination in spite of their unhelpful genetic blueprint.

No one can yet be sure why some people put on weight while others on a similar regimen do not. All we can say is that in our own species the mechanism for controlling weight is unreliable in its operation, in much the same way as the mechanisms controlling salt balance, and regulating temperature. In these respects our bodies do not respond as promptly and as appropriately to our present-day needs as the bodies of most animals do.

Another penalty we pay for the possession of our subcutaneous fat layer was described in 1955 by Peter Medawar who referred to it as '. . . the appalling ineptitude of wound healing in the human skin'.

In mammals generally the skin is very loosely connected to the body wall. Anyone patting a spaniel is aware of how easily its coat slides to and fro over the underlying flesh. Mammals also have muscles in the skin so that they can move it independently, as a horse does when it twitches its skin while pestered with flies. Humans have almost entirely lost these intrinsic skin muscles — except in the face where they enable us to change our expression.

When tom cats retire to lick their wounds after a fight, they may have numerous lesions in the skin, but, once the bleeding has stopped, the edges of the cuts lie side by side, a thin film of skin epithelial cells migrates inwards and the skin edges are drawn together by a process of contracture until they meet. Healing occurs very quickly, and without permanent scarring. In a rabbit, for example, an area of missing skin up to a hundred square centimetres, which in humans would necessitate a skin graft, will be naturally repaired and heal so that the scar is almost invisible.

In humans the skin is bonded to the layer of fat, which often prevents the edges of a cut from uniting. We speak of a man's cheek being 'laid open' or of a 'gaping wound' in the thigh, and the thicker the fat the wider the gape. This is the phenomenon which impressed Professor Wood Jones when he dissected cadavers, and Alister Hardy when he watched a seal being cut up. To many a surgeon obliged to operate on an obese patient, the comparison with blubber must seem uncomfortably apt. Medawar comments:

> The upshot of this new anatomical arrangement is that contracture, far from being an efficient mechanism of wound closure, has become something of a menace; it constricts, disfigures and distorts, and yet

may very well fail to bring the edges of the wound together. As a mechanism of wound healing, the contracture of human skin is therefore as archaic as the vermiform appendix. We only become aware of it when it leads to harm.

Nowadays we are seldom aware of it because it is standard practice to go to a doctor to have some stitches put in or, in severe cases, for a skin graft, but for primitive man an untreated wound could cause a seriously disabling injury by failing to heal, or by gathering up and scarring in such a way as to constrict the blood supply or immobilise a limb. Marine animals would seem to suffer similar disadvantage. Manatees in Florida have had their habitats invaded by pleasure boats, and their skins are sometimes sliced into by the blades of underwater propellers. The ones who manage to survive the experience show, not the narrow trace of a healed cut, but a broad white scar where the pigmented outer layer of the skin has failed to come together.

Medawar's final comments strike a familiar note: 'What compensatory advantage the human being gets from the novel structure of his skin is far from obvious, though it is hard to believe there is none.' The compensations on land are indeed obscure, but for aquatic mammals at least two possible advantages of the fat layer can be readily detected — insulation and buoyancy.

The subcutaneous fat layer, while in air it provides less efficient insulation than a coat of fur, is extremely efficient as a protection against heat loss in water. P.F. Scholander and his colleagues published a study of body insulation in some arctic mammals in 1950. They studied a seal (*Phoca hispida*) and a polar bear. The seal, which spends most of its time in the water, has a thin hair covering and a thick layer of blubber. The bear, which spends most of its time on land, has a thick fur coat but, except when preparing for hibernation, not much fat under its skin. When the polar bear moved from zero degree air into the water, it lost body heat 50 per cent more rapidly than on land. When the seal did the same, it lost heat only five per cent more rapidly than on land.

In order to serve this purpose of insulation in water, the adipose tissue in aquatic animals not only constitutes a higher proportion of the body tissue, it is also differently distributed. In aquatic species, by comparison with land animals, the tendency is for fat at the internal sites — around the kidneys and intestines — to be diminished, while the superficial (that is, subcutaneous) fat deposits are disproportionately enlarged. For example, in some land mammals such as the horse, 50 per cent of the fat in the body is found in the abdomen. In some species of seal the fat at this particular site — the mesenteral deposit — has virtually vanished, whereas the blubber under the skin is very thick.

In humans the redistribution has not gone as far as in whales and seals, but it appears to have moved quite a long way in that direction. We retain a supply of internal adipose tissue, but the subcutaneous deposits are greatly expanded. In species like cats and dogs the 'fat' under the skin amounts to little more than a thin film of cells serving as a lubricant, enabling the skin to slide easily over the muscular tissues beneath. Thus, an attacking enemy or predator is more liable to end up with a mouthful of fur and skin rather than inflicting deeper damage.

In humans, by contrast, it has been calculated that on average, between 30 and 40 per cent of all the fat in our bodies is located just under the skin. Caroline Pond, who has added greatly to our understanding of the anatomical distribution of adipose tissues in mammals, once wrote: 'It is difficult to see what mechanical or thermoregulatory function the subcutaneous fat on human limbs could serve. By increasing the inertia of the limb, it could actually hinder movement.' In water, as an insulator, it would make sense.

Another advantage of fat to an aquatic animal is that it gives buoyancy. Of our body tissue, fat and lean have different densities, so that while they weigh the same in air, they do not weigh the same in water. Drop a piece of lean meat into water and it sinks; a piece of fat will float. Swimmers are not among those athletes who strive to shed every surplus pound in order to excel at their particular sport. Among aquatic mammals it is notable that a surface feeder like the white whale has fifty times as much blubber as it would need solely for insulation. The walrus does not need so much buoyancy because it gets its food from the sea bed, so though it lives in the same chilly latitudes its fat layer is much thinner, and varies with the seasons and the food supply. That is why a dead whale floats and a dead walrus sinks.

Thus, the same fat which could be a burden to a land ape, hindering locomotion and incurring extra energy costs, would be a benign development in an aquatic one. It would convey a saving of energy costs both by reducing heat loss and by enabling the animal to stay afloat more effortlessly.

10

Explaining the Fat Layer

'Land dwellers have relatively thin skins.
The subcutaneous fat between the subcutaneous
muscles and the muscles of the
body is rarely a continuous thick layer.'
V.E. Sokolov

Those for whom the aquatic hypothesis is unacceptable have to seek some other way of dealing with the problem. The simplest ploy is to ignore its existence. The two textbooks on physical anthropology, mentioned at the beginning of Chapter 6, which between them devote three words to the loss of body hair, devote none at all to subcutaneous fat. It is another illustration of Medawar's dictum that scientists tend not to ask themselves a question until they have a glimmering of the answer.

One suggestion has been that the fat layer evolved as a way of storing energy, and this theory is supported by citing other animals which accumulate fat under the skin, such as bears, marmots, hedgehogs and dormice. These are all hibernators; the fat layer is seasonal; they all belong to latitudes where winter brings cold and scarcity of food. In some warmer latitudes where an animal needs to store energy it stores it in some special depot where it will least impede locomotion; examples include the camel's hump and the tails of the fat-tailed sheep and the fat-tailed gecko. None of these features applies to *Homo*.

Above all, the storage theory does not explain why only

114

one primate species on the savannah needed to store energy, and it does not explain why the fat layer is thickest in babies. At that stage of their development, when they were being fed by their mothers, the danger of suffering a food shortage surely could not have been greater for hominid babies than for the young of other apes.

In the context of the savannah theory, the baby's fat is in some respects the most baffling of all. It might be argued that a hairless baby has special need to guard against sudden chills. But a closer examination indicates that this cannot be the explanation.

Fat comes in two kinds — brown fat and white fat. Brown fat (it is the darker in colour because it has a much larger blood supply) is a special type found in all young mammals, and its function is to supply heat quickly. It is necessary because when they are first born they are unable to raise their temperature by shivering. Brown fat responds to a drop in body temperature by burning up quickly and converting into heat. White fat is predominant in mature mammals. It is stabler, does not burn up easily, and does not respond to a drop in temperature. White fat cells are depleted only when the body's intake of calories is insufficient to replace the energy which has been expended.

If our babies became fat because being naked made them liable to chilling, we would expect most of their extra adipocytes to take the form of additional brown fat. The reverse is the case. They have the normal allowance of brown fat, but they are extremely unusual in also possessing appreciable amounts of white fat at birth. By the time they are three or four months old, all their brown fat has already been converted into the mature white form — good for insulation in water and for buoyancy, but no good at quickly raising body temperature.

More popular than the storage idea is the theory that is currently still the front runner. It proposes that the evolution of a fat-lined skin was part of a multi-stage strategy for regulating temperature. The story goes that the ancestral ape became a hunter and his exertions made him too hot; so he lost his fur to cool down; he then became too cold, especially at night, and installed a layer of fat to keep him warm; the fat made him too hot, so he evolved the capacity for profuse sweating to keep him cool.

There are variations on the theme. In some versions he was not a hunter but a gatherer; in some accounts the sweat came first, with the nakedness following to enable the sweat to evaporate. But none of the versions has produced a satisfactory explanation of why one primate species alone needed to resort to these complex ways of replacing a thermoregulatory system which works perfectly well for the other savannah animals, including savannah primates.

All in all, the concept of fat hominids evolving on the savannah is so inherently improbable that some attempt has been made to side-step it. This is done by suggesting that the subcutaneous fat layer did not evolve on the savannah, but only appeared at a much later date, in an agricultural economy. It is reasoned that the carbohydrate intake of agricultural man would be higher, and he would not need to race across the plains in pursuit of his dinner, so he could afford to carry a little more weight.

On the other hand, he would have even less need to evolve ways of storing surplus nourishment under his skin if he already had ways of storing it in barns and receptacles. Moreover, there are extant tribes who have never practised an agricultural economy and yet their young women have rounded breasts and limbs and chubby infants like the rest of humanity. And the 'Venus' figurines of women with fat thighs, buttocks and bellies predate the domestication of plants and animals.

The last of the non-aquatic hypotheses about fat suggests that the fat evolved first in the females to make them more easily distinguishable from the males and more attractive to them. The females then presumably passed on the new attraction to their sons as well as their daughters, though in a modified and more meagre version.

Undoubtedly there are wide differences between the sexes both in the amount and distribution of their adipose tissue. The percentage of fat in a woman's body is on average about twice as high as in a man's. And apart from the characteristic sex-linked accumulations on breast and buttocks, the fat on females tends to be more evenly distributed. Overweight women tend to be fat all over, while overweight men normally develop 'male pattern obesity'. This is the condition where the fat accumulates disproportionately in the abdomen, giving rise to 'the beer belly' silhouette. It is often combined, as in Dickens's Mr Pickwick, with quite thin arms and legs.

But there is one strong argument against the subcutaneous fat deposit having evolved initially as an epigamic marker — that is, for sexual attraction. The whole point about epigamic markers is that they are a sign not only of sexual differences but of sexual maturity. Young animals do not acquire manes or antlers or breeding plumage; these things only arrive with puberty. There is no conceivable reason why human infants should have such generous deposits of fat all over their bodies if the original purpose was that of arousing the sexual interest of the male.

As for the female's extra ration of fat, the explanation of that is more likely to be practical than erotic. The gestation and breast-feeding of a human baby make heavy demands on the mother's physical resources. If at any time during that period her intake of food is insufficient to meet those needs the white fat cells in her own body are drawn on to ensure that the baby is not affected.

For hominids in the wild, the bearing and feeding of offspring would not have been an occasional event in the life of the female. In laboratory apes the infant is generally removed from the mother at the end of a year, or at most two years, but in the wild, breast-feeding continues longer. When an infant is weaned — or if by chance it dies — the oestrus cycle soon recommences. But it does not last for long. On average only between three and four cycles elapse before the female is again pregnant. With the exception of these brief intervals, for the greater part of her life the physical needs of a bouncing baby would have imposed on a hominid mother an ongoing need for reserves of white fat which the male did not share.

In apes, the non-stop cycle of pregnancy and feeding continues till death; in human females it ends with the menopause. No one is quite sure why women are granted this dispensation, which is very rare in nature — though it has recently been claimed that we share it with whales and dolphins.

One suggestion is that the menopause tends to prolong the active life of grandmothers by lessening the physical strain on them. The young of humans take a very long time to mature, and the success of our species depends to a unique extent on the passing on of acquired knowledge and skills to the next generation. So it would have been advantageous for a breeding population of hominids if there were some experienced non-breeding females present to help in the task of child rearing.

It seems a little odd that just around the time of the menopause, when there are not going to be any more children to gestate and suckle, there is a general tendency in women of all races for their weight to go up instead of down, and for an increased percentage of their body tissues to consist of fat.

But it has now been learned that adipose tissue performs additional functions besides insulation and buoyancy. Besides being constantly active in the turnover of fuels in the body, it is capable of storing steroids (hormones), and it influences both the amount and the potency of the female sex hormone oestrogen circulating in the blood.

Oestrogen occurs in two forms — a relatively inactive form and a highly potent form. Thin women have elevated levels of the weak type, while fatter women have higher levels of the strong. In fatter women the oestrogen also circulates more freely in the blood stream. Until 1975 it was believed that oestrogen was produced only in the ovaries; it was then discovered that it is also produced in fat cells, by converting the male hormone androgen (found at low levels in female blood plasma) into oestrogen.

It is not inconceivable that the 'middle age spread' which so many women battle against evolved as a benign development. It would ensure that the body's auxiliary source of oestrogen (adipose tissue) was in plentiful supply. When the ovaries ceased to function this would minimise the withdrawal symptoms. It would enable the body to adjust more gradually to the drop in oestrogen levels instead of quitting, as it were, 'cold turkey'.

For most people in the developed countries today it is easier than ever before for people to eat and drink too much and take too little exercise. In these conditions the capacity of the human body to gain weight rapidly and far in excess of what is needed constitutes a problem. Because it can have adverse effects on health and morale, interest in the subject has grown rapidly in recent years.

One good thing has come out of the contemporary obsession with slimming. It has prompted active research into a subject which most scientists have previously found boring. In the '70s a textbook on animal anatomy, over 500 pages long, devoted only one page to a description of adipose tissue before concluding with the comment that '. . . little more seems necessary to be said about the subject'.

Scientists wishing to find more to say about it were confronted with an initial difficulty: most land mammals, especially those suitable for laboratory observation, do not resemble humans in the distribution of their adipose tissue. In a normal rat, for example, only about two per cent of the body consists of fat tissue, and however much it is overfed, it fails to become obese.

For a considerable period, therefore, research concentrated disproportionately on one aberrant strain known as the 'obese rat'; and it was thought that lessons learned from an abnormal rat could be applied to perfectly normal humans. It was found, for example, that if obese rats under the age of three months were allowed more than their share of the maternal milk supply, or weaned earlier onto solid food, they grew extra adipocytes and ended up fatter than the other rats from the same litter. Women were given stern warnings about overfeeding their babies or introducing them to solid food prematurely. Plump teenagers cast reproachful eyes on guilty mothers and blamed them for their unhappy plight.

Later it was discovered that data on obese rats did not hold true for human babies, and amendments were issued on the lines of: 'The early introduction of solid or bottle feeding does not increase the risk of infant obesity. Anxious parents should be given reassurance on this point.'

Clearly, people are a special case. The trouble with using them for research is that it often calls for long-term planning and the results are slow to emerge. There was a general agreement among doctors that people who are significantly overweight have an increased risk of becoming ill or dying, especially from cardiovascular disease — heart attacks and strokes. So it seemed only sensible to conclude that the fatter people got, the more likely it was that they would suffer from these conditions, and *vice versa*.

This belief was not easy to verify. But one of the functions of science is to convert facts that we are nearly sure we know into facts that have been checked and proved. So in 1967 in Germany a project was set up involving 782 men of the same age (54 years old) resident in Gothenburg and selected on a random basis. Records were kept, including measurements of height, weight, blood pressure, cholesterol levels and the circumference of waist and hips. There was a hundred per cent follow-up on the subsequent medical histories of the men over the next thirteen years, with repeat examinations in 1973 and 1980.

The records showed that the group with the lowest risk of premature death or coronary heart disease were the fattest men, as calculated by weight in relation to height (The Body Mass Index). The highest risk group were in the leanest category by the same height-weight ratio. It was precisely the opposite of the result that the researchers had expected.

These findings only began to make sense when the figures were analysed in greater detail. It was then found that the crucial factor was not the overall amount of body fat, but the way in which it was distributed. The high risk group consisted of men with moderately lean legs and buttocks but with body fat concentrated in the abdominal region — the type of distribution identified by the French physiologist Vague in the '60s as 'android' or male type. The low risk group of men weighed the heaviest, but their fat was more evenly distributed all over the body in the gynoid (female type) pattern.

Similar studies were subsequently carried out on women in Germany and France and the results indicated that the same general rules applied. Women with more android figures were more at risk than fatter women with a more gynoid silhouette.

Explanations of why a fat belly is more unhealthy than fat thighs and buttocks are still tentative. One idea is that free fatty acids from adipocytes in the abdomen go directly to the liver via the hepatic portal vein and overload it, while fatty acids from the thighs travel around the body and reach the liver by a more roundabout route, and by the time they reach it they are in lower concentrations.

The lesson to be learned from the Gothenburg experiment would seem to be that — for women, at least — a few extra pounds are not as serious a health risk as previously supposed. In Australia not long ago a young couple in their thirties, Michael and Sue Murnane, in all other respects seen as potentially 'perfect parents', were forbidden to adopt a Korean baby on the grounds that they were overweight. In Korea and Sri Lanka the law apparently assumed that any adopters weighing 30 per cent more than the average would be liable to leave the baby orphaned before it had time to grow up. That seems an unnecessarily alarmist proposition.

Nowadays scientists are only responsible for a tiny percentage of the millions of words of advice and exhortation which pour from the presses on this subject. Helping people to lose weight has become a multi-million dollar industry with a massive advertising budget. Insofar as it encourages people to eat less and eat more sensibly, it is doing an excellent job and can take credit for improving health awareness and the general standard of fitness.

But there is another side of the coin. The slimming gurus, like other glamour promoters, are in the business of selling dreams. When they prescribe their norms of the 'correct' vital statistics for a particular age group to aim at, they have every financial incentive to pitch the figures low. The 'target weights' are lower than the average; lower than most people can easily sustain, and lower than health considerations alone would dictate. There is even some reason to believe that they are lower than considerations of sex appeal would dictate. In one experiment, a panel of males were shown pictures of women ranging by gradations from thin to fat, and asked to pick out the most attractive. A panel of women was then invited to guess the men's verdict. The results consistently indicated that men are less attracted to a Twiggy-type female silhouette than women imagine they are.

The subliminal message in some of the 'advice' commercially offered is that the average female is too fat, that fat means ugly, that failure to get rid of it is something to be ashamed of. It lends credence to the Duchess of Windsor's dotty dictum: 'You can never be too rich — nor too thin.' It may seem strange that in our day physically normal healthy people will pay surgeons to operate on them to bring their figures closer to an arbitrarily prescribed ideal. But it is not a new phenomenon. The same impulse once drove Chinese ladies to resort to the service of a foot binder, and English ones to demand surgery on their lower ribs to help them attain a sixteen-inch waist.

The real casualties of the 'never-too-thin' propaganda are for the most part young people. Some of them believe the message, and it can induce in them a debilitating loss of self-esteem leading to depression or a flight into food obsessions like anorexia.

In 1985 the gynaecologist John Studd conducted an investigation into the immediate and more long term effects of anorexia. He reported: 'Oestrogen isn't only produced in the ovaries. It is partly made in body fat. The less you have, the less you make . . . It isn't uncommon to find an anorexic woman of 28 or 35 with the skin and bones of a 70-year-old.'

Research based on his findings suggested a correlation between low oestrogen and reduced levels of collagen. Collagen is the protein which constitutes one-third of the tissue of bones and skin. In some of his young patients Studd found not only a loss of skin tone (reversible when normal eating habits are restored) but a permanent loss of height through subsidence of the vertebrae due to reduced levels of collagen.

There has to be some evolutionary reason why *Homo sapiens* is the fattest as well as the sweatiest of all the apes. It is bad luck that our bodies have lost the capacity to impose an appropriate upper limit on the amount of fat they can accumulate. The problem we face is how to avoid obesity without letting the fear of it escalate into a phobia, since adipose tissue remains an indispensable part of our anatomy.

11

Breathing

'It is a strange fact that every particle
of food and drink we swallow has to pass
over the orifice of the trachea, with some
risk of falling into the lungs.'

Charles Darwin

Darwin was not overstating the matter. It is indeed a
strange fact, which looks like a deplorably clumsy piece
of biological design — all the stranger because it is another
of the many ways in which humans differ from all other
land mammals.

The basic mammalian design is one in which air enters
through the nostrils and goes down through one tube into
the lungs, while food and drink enter through the mouth
and go down through another tube into the stomach. The
tube for transporting air is the windpipe (trachea) and
the tube for transporting food and drink is the gullet
(oesophagus). In most mammals it is impossible for food
or drink to 'go down the wrong way' into the lungs,
because the upper opening of the windpipe is not situated
in the mouth, but higher up, at the back of the nasal
cavity. Everything the animal eats or drinks has to pass
around or to one side of the upper section of the windpipe
before reaching the gullet and descending into the
stomach.

It is a simple and foolproof system which keeps the
openings of the two channels entirely separate. Anyone

who has watched a thirsty horse drink (or especially a thirsty camel) will have noticed that the animal can go on drinking for a long time without having to pause to draw breath. Because the air and food channels are separate all the way, they can breathe and drink simultaneously.

In all land mammals except man, the top of the windpipe is in contact with the palate, which is the partition separating the mouth cavity from the air passages. The front part of it is a bony partition, the roof of the mouth. Behind that in most animals the palate is extended backwards in a continuous sheet of tissue known as the soft palate. In primitive animals, such as some reptiles and a few present-day marsupials, the top of the windpipe (the larynx) passes through the soft palate and is permanently locked into position there. These animals are therefore incapable of breathing through their mouths, and they cannot utter any vocal sounds.

In the majority of mammals nose-breathing is still the normal procedure, but the barrier against mouth-breathing is less absolute. The larynx is not permanently locked into position: it passes through a small aperture in the soft palate which is controlled by a sphincter — that is, a contractable ring of muscle. A dog, for example, may want to breathe through its mouth temporarily for the purpose of panting to cool down, or in order to bark. When that happens, the sphincter muscle is relaxed and the top of the windpipe slides down through the hole in the soft palate, just far enough to bring the larynx down into the mouth cavity. When a dog barks or howls it often stretches its neck, and that helps to disengage the larynx from the palate. When the panting or howling is over, the neck relaxes, the larynx goes back up through the hole, the sphincter tightens around it, and the dog is once again a nose-breather. Some version of this arrangement holds true for most mammals.

In adult humans the larynx has lost all connection with the palate. It has slid right down into the throat, below the back of the tongue, so that, as Darwin pointed out, the opening of the windpipe is in front of the gullet opening and on the same level. Our soft palate is no longer a continuous sheet of tissue with a neat, closable hole in it to accommodate the larynx. There is a large gap in it. The uvula, the conical fleshy prolongation which can be seen dangling at the back of the throat, represents part of the upper rim of the hole. We can therefore breathe with equal facility either through our noses or through our mouths. But we cannot breathe and drink at the same time, and if anything makes us gasp while we are drinking, some of the liquid is inhaled into the lungs.

This evolutionary change is known as 'the descended larynx', and much scholarly speculation has gone into trying to work out why it is an improvement on the normal arrangement. (There is always a tacit reluctance to believe that any anomalous features characteristic of our species can be other than an improvement.)

Perhaps the most unexpected eulogy of the descended larynx came from an enthusiastic surgeon who called it an excellent arrangement, because human patients with injuries to the mouth cavity can be easily fed through a tube passing into the stomach via the nose. That is indeed one of the incidental benefits of laryngeal descent, but it can hardly explain why it came about.

Like the spine of a quadruped, the normal mammalian blueprint for breathing has been perfected over more than two hundred million years and would not have been abandoned for any trivial reason. It might be as well to count up the advantages we have lost before searching for what we have gained.

Nose-breathing is important to most mammals on five counts. Firstly, the sense of smell is of vital importance to them, and the airborne pheromones can only be detected if the air passes over the scent-detecting organs in the nasal passages. Secondly, nose-breathing protects the lungs by filtering out noxious substances. Particles of airborne dust are trapped by the hairs lining the nostrils. Finer particles are also arrested before reaching the lungs; they adhere to the mucus-lined passages of the nostrils and the nasal passages into which they lead, and are eliminated by one of a variety of routes — sneezing, nose-blowing, spitting, or being swallowed with the saliva.

Thirdly, air which travels through the nasal passages is either warmed or cooled to near body temperature before it reaches the sensitive and vulnerable tissues of the lungs. Fourthly, if the air is arid, it is moistened by the secretions of the mucous membranes. Finally, some of the mucous secretions are mildly bactericidal. Air inhaled through the nostrils reaches the lungs filtered clean, warm, moist and to some extent sterilised.

All these safeguards are discarded when we breathe through our mouths, as all of us do for part of the time — for instance, when we are talking, running, or doing any task involving physical exertion.

The lungs of all mammals have some ability to dislodge — by coughing — any noxious particles which somehow penetrate the outer defences, but in most mammals this is a last-ditch emergency procedure; lungs did not originally evolve to cope with unprocessed air. Our pets breathe the same air as we do, but they do a lot less coughing. The ones most troubled with breathing disorders are breeds like Pekinese dogs, in which selective breeding has artificially reduced the length and efficiency of the nasal passages.

The descent of the larynx entailed a whole sequence of modifications in the upper respiratory tract. The consequences of these changes are sometimes inconvenient, and occasionally lethal. One of them concerns what happened to the tongue.

Swallowing is a uniquely complicated process in humans because of the need to prevent food and drink from entering the windpipe. The device which has evolved to ensure this is as follows. The back of the tongue, unlike that of most mammals, projects an appreciable distance beyond the back teeth. Every time we swallow food or a drink of water, or even our own saliva, the top of the windpipe moves upwards in the throat and buries its head (the larynx) underneath the back of the tongue. This movement forces a flap of cartilage (the epiglottis) to bend over and form a lid over the larynx which shelters it from the incoming liquid. When the liquid has passed safely down into the stomach, the windpipe drops back down to its normal site.

If the swallower is an adult male, this manoeuvre is usually visible to the naked eye, due to the fact that in males around the age of twenty, a ridge of cartilage at the front of the larynx begins to be replaced by bone. In some men this bony projection, known as the Adam's apple, is a prominent feature and can be seen bobbing up and down with every swallow.

The backward projection of the tongue does not, of course, reach back far enough to block the airway — as long as we are conscious, perpendicular, and young. But with increasing age there is often a deterioration of muscle tone. Then, for a mouth-breather who falls asleep with the mouth open, problems can arise. The tongue may slump farther into the back of the throat by the mere force of gravity, so that the air has to force its way through the diminished opening with the noisy rattle known as snoring. If the sleeper's airway closes up entirely as it sometimes does, then the lack of oxygen in the blood stream alerts the body to its danger, and the mind emerges from its slumber at least far enough to ensure that the blockage is removed and free breathing is restored.

However, if a similar blockage takes place during deeper unconsciousness, as for example in the case of concussion following an accident, it may be fatal. It is because of this uniquely human hazard that instructions for First Aid to accident victims begin: 'First check the airway.' Deaths caused in this way, like the occasional death of an alcoholic asphyxiated by his own vomit, are among the hazards incurred by having a descended larynx.

Within the last decade publicity has been given to a condition which has been named 'obstructive sleep apnoea' ('apnoea' means 'without breath') or OSA. Normally a sleeper, when suffering a blockage of the airway, first tries to breathe in spite of the blockage, with spasmodic heavings of the rib-cage and diaphragm. This in itself can have deleterious effects, especially for sufferers from heart disease, by causing abnormal changes to air pressure in the lungs, increased blood pressure in the chest and decreased level of oxygen in the blood. However, it usually only lasts for about fifteen seconds before these symptoms cause a partial return to wakefulness, whereupon normal breathing is restored, and sleep is resumed. The cycle usually repeats itself many times, in severe cases several hundred times in a night.

In North America there are now numerous clinics specialising in sleep disorders, and doctors have found a correlation between OSA and hypertension (high blood pressure). OSA is not a primary cause of hypertension but it may be a contributory or aggravating factor. According to one theory the correlation only means that both conditions result from some other cause such as, for example, being overweight. But a leading researcher, Christian Guilleminault, reports that only half of those who suffer from OSA are overweight, and hypertension still occurs in the lean OSA patients.

The condition can cause psychological as well as physical ill-effects. Episodes of apnoea are normally commoner and more severe during the REM (rapid eye movement) phases of sleep when dreaming takes place; these are the periods of deepest unconsciousness. According to current thinking, they are also the phases we can least afford to be deprived of.

When a patient's REM sleep is disturbed by frequent

episodes of OSA — even though he does not remember them in the morning — it may result in persistent sleepiness and fatigue during waking hours. Also there are often complaints of deterioration of memory and judgement, early morning confusion and headaches, and attacks of irrational anxiety and depression. The symptoms resemble in a milder form the effects induced in prisoners who are deliberately deprived of sleep while being interrogated. The recommended treatment varies from prescribing weight loss and the avoidance of alcohol late in the day to special appliances pumping air through the nose, or surgery to remove part of the soft palate and tighten the muscles in the throat.

No baby is born with a descended larynx. For the first few months of life its respiratory passages are cast in the normal mammalian mould, with the larynx high up, in contact with the palate and in communication with the nasal passages. It is a habitual nose-breather. When a young baby suckles it can suck and breathe at the same time, and the milk flows steadily down into the stomach without the complicated swallowing procedures necessary in adult life.

Some time between the third and sixth months after birth, the larynx loses contact with the palate and begins to descend. There is a transitional period during which the baby's upper respiratory tract is not quite animal and not quite human, and its reflexes are in the process of learning to adapt to its new physiology.

This period coincides with the peak incidence of Sudden Infant Death Syndrome (SIDS). Unexplained deaths in this age group used to be accounted for by saying that the baby had been 'overlain', and mothers were instructed that it was safer and healthier for babies to sleep alone. Now that babies do sleep alone, these fatalities are called 'cot deaths'.

Usually post mortems reveal nothing about the cause of death. Records show that the incidence of cot deaths is somewhat higher in winter. Correlation with recorded weather reports shows that the highest peaks occur not during but from two to five days after sudden spells of very cold weather, during which the baby's resistance to virus infection may have been lowered. This is consistent with the possibility of an infection of the upper respiratory tract — in other words, the baby has caught a cold. But this does not explain why death should be caused — in this age group specifically — by an infection so mild that in the great majority of cases the parents have not noticed it, and the autopsy reveals nothing.

The possible connection with the migration of the larynx was first perceived by Edward Crelin who discovered over twenty years ago, when about to operate on a baby, that there was no textbook on infant anatomy to which he could refer for guidance. Clinicians worked on the assumption that human babies were constructed like miniature adults. So Crelin set about compiling a book on the subject which is now used as a work of reference in hospitals all over the world. One of the major discoveries he made during his research for this project concerned the way babies breathe. It was a complete surprise to him. He had expected the baby's anatomy to resemble a miniature version of a human adult's anatomy, but when he came to the upper respiratory tract he found himself 'dissecting the airways of a chimpanzee'.

When the larynx first begins to descend, although physically it is becoming possible for the baby to breathe through its mouth, its reflexes are still those of a nose-breather. If the nasal passages are blocked because of an infection, it shows distress and may cry out. As in any other animal, vocalisation detaches the larynx from the palate and enables the baby to breathe, and it may drop off to sleep again with the mouth open.

The risk arises if it is sleeping on its stomach, or on its back with the head low. The larynx may then slide back up to a position where the uvula can enter it and block it. After death, the muscles relax, the larynx disengages when the baby is picked up, and there is nothing to show the cause of death except perhaps some slight damage to the lung.

Crelin believes that 90 per cent of cot deaths are caused in this way. His theory remains a theory, but it has been reinforced by clinical experience. For example, a Midwest paediatrician, the late Harvey Kravitz, advised mothers to raise the head end of the cot a little, and to keep the baby's head up when out of the cot; in the next twelve years there was not a single cot death among the 1,800 babies he dealt with, whereas statistically about five would have been expected.

A report from the Netherlands involved much larger numbers. In the early 1970s Dutch mothers were advised to place their babies face down to sleep, and by the 1980s at least 60 per cent of them were following this advice. During that time the incidence of cot deaths in the country went up from 0.46 per thousand births in 1969–71 to 1.13 in 1987. In 1987 well-baby clinics throughout the country began to advise against babies sleeping face down and a 40 per cent decrease in cot deaths was recorded between 1987 and 1988.

These statistics, plus the fact that cot deaths are very rare before the age of three months and after the age of six months, are consistent with Crelin's belief that they are connected with the transition from the infantile to the adult-human location of the larynx.

Early colds and infections may cause trouble later on for children who do not succumb to cot deaths. If they are forced into mouth-breathing too early because of nasal obstruction, they sometimes acquire the habit of habitual mouth-breathing, known to doctors as HMB, and this leads to a kind of vicious circle.

Because the nasal passages are being by-passed and not ventilated, secretions accumulate there which further obstruct the air flow; dried mucus forms a perfect breeding ground for micro-organisms which can result in inflammation and overgrowth of the adenoid tissue. Open-mouth breathing also leads to a higher evaporation rate of saliva, so that less of it needs to be swallowed. Normally, frequent swallowing helps to clear and ventilate the middle ear via the eustachian tube. A Netherlands team (Van Bon *et al*) in 1989 confirmed the hypothesis that HMB may be a cause of otitis media (earache) in young children. Early in this century, descriptions of slack-jawed, snotty-nosed slum children were sometimes taken as signs that they were hereditarily subnormal, rather than as a condemnation of the overcrowding in cold, damp conditions which spread the infections in the first place.

Understanding of the evolution of the larynx grew slowly, and one reason was that anatomists study corpses and not living animals. Drawings of dissected animals by different anatomists showed the larynx in different positions, and led to heated controversy in the 1880s.

It was finally realised that the different versions depended partly on how the animal died — if it died crying out, then the larynx would be found in the mouth cavity — and partly on how the body was manipulated during dissection. Once the sphincter is relaxed, simply stretching out the animal's head and neck may cause the larynx to slide easily out of the nasal chambers and thus be found in a more human-like position.

The matter was finally clarified by an examination of live and conscious animals — often by subordinates rather than the anatomists themselves. Dr R.L. Bowles in 1889 wrote of one of these unsung heroes: 'My assistant Mr Stainer has introduced his hand into the pharynx [back of the throat] of many pigs of six months and upwards.' Stainer found that in the living animals the windpipe passed up through the soft palate. It was held in place there by the epiglottis (a small flap attached to the top of the larynx) with the sphincter muscle of the opening in the soft palate 'firmly clasping its root'.

Nowadays there is agreement that the descended larynx in humans is a rare anomalous feature, but there is no agreement as to why it evolved. Sir Victor Negus spent his lifetime researching the anatomy of the larynx and he was in no doubt about whether or not man stood to gain from this odd change in his anatomy. It was, Negus declared, 'greatly to his disadvantage'.

He never found any convincing explanation of why it had happened. He mentioned that man was micro-smatic — that is, with a diminished sense of smell — so to that extent he had slightly less to lose by ceasing to be a habitual nose-breather. He noted that the tongue 'extends for some considerable distance beyond the last tooth', and speculated whether 'the larynx may be said to be pushed down the neck by the tongue'. (But unless and until the larynx had descended, there was no reason why the tongue should have acquired its backward projec-tion.)

Finally he leaned towards the theory still in favour today — namely, that the descent of the larynx was in no way adaptive; it was simply one of the many unfortunate unavoidable side-effects of bipedalism. Bipedalism necessitated changing the angle at which the head is set on the spinal column, so that the human gaze is directed forward at right angles to the spine, instead of in the same plane as in quadrupeds. It is argued that this change in angle caused the windpipe to slide down the throat, aided possibly by the force of gravity.

One exposition of this theory points out that '. . . in mammals where the larynx is placed high in the neck such as cats, dogs, monkeys and apes, the basicranium is relatively non-flexed' — that is, the head is at the normal angle for quadrupeds. The argument sounds persuasive until you appreciate that the relatively non-flexed category includes not only cats, dogs, monkeys and apes, but every terrestrial mammal except man.

There are three reasons for rejecting the 'unfortunate

side-effect' theory. One is that in a foetus or a new-born human the angle of the head is even farther removed from the quadrupedal angle than in an adult, yet the larynx is placed high, as in animals. Secondly, experiments by Jan Wind in the Netherlands have established that gravity does not affect the issue. Certainly, when a baby's larynx descends the baby is still in the horizontal phase of its life, so gravity can hardly be pulling it down.

But the strongest argument against the theory was pointed out to Negus by his greatest admirer and supporter, Sir Arthur Keith. If there had been any tendency to slide, only a slight modification would have been needed to counteract it, and '. . . if Nature had any necessity for maintaining close relations between nose and larynx, it would find means to do so.'

When Negus spoke of 'great disadvantage', he was thinking of disadvantage to a land animal, for which an uninterrupted passage from nostrils to lungs is a high priority — provided that what enters the nostrils is going to be air. For an animal that spends much of its time in water, the priorities are somewhat different.

One priority for swimmers and divers is the ability rapidly to inhale and/or exhale large quantities of air, either before diving or when briefly surfacing. This necessitates a wide opening. Negus made this point in connection with the wide laryngeal opening of the horse. 'If any animal has the necessity of supplying a large volume of air to its lungs, it must have a wide tract, and cannot attain the required result by mere increase in the rate of respiration.'

When we breathe through our nostrils, our own air tract is not merely narrow but — compared with that of a quadruped — also tortuous. In a dog, air travels in a straight horizontal line from nostrils to lungs; in man it goes up, and across, and down. By far the quickest way of increasing the volume inhaled is to breathe through the mouth.

The other requirement for aquatic species is to be able to close the air passages. Any mammal accidentally falling into the water automatically holds its breath by closing the glottis — the space between the vocal cords — and inhibiting respiratory movements of the lungs. This is adequate as an emergency procedure, and most terrestrials do not need to sustain the closure for long, because their mode of swimming, like the dog-paddle, is designed to keep the head above water at all times.

Diving species need more protection. A method many of them have evolved is that of acquiring valvular nostrils like those of a seal, which in a relaxed state are closed, but can be opened at will on surfacing. Aquatic species which are unable to use this method resort to other devices. In birds, for example, the nostrils are openings in the beak, and they cannot be closed. The penguin therefore has a triangular flap of cartilage inside the windpipe which has, like our own larynx, lost all connection with the nasal passages. Like many other aquatic birds, such as pelicans and gannets, the penguin has become a mouth-breather.

Several unrelated aquatic species have evolved some kind of internal movable flap either instead of, or in addition to, valvular nostrils. The penguin has one; and the crocodile has one. Alone among the primates, humans have such a flap — that is, the back of the soft palate, known as the velum, which in our species can be raised and lowered to isolate the nasal passages from the mouth cavity. It could not operate in this fashion if the larynx had not retreated out of its way to its present position below the back of the tongue.

The only other mammals which are known to feature a descended larynx are diving mammals — the sea lion and the dugong. These two species are as about unrelated to one another as they are to humans. The descended larynx must have evolved independently in each of them, after their respective land-dwelling ancestors entered an aquatic environment.

If these unusual features of the human respiratory tract were indeed aquatic adaptations, by now their efficiency has been in some respects eroded. There may have been a time when the velum closure was more powerful than it is today. There may have been a time when the wings of our nostrils (not found in other primates, and muscled in the same way as a seal's) were fully vascular and could open and close.

The descended larynx, however, has stayed with us permanently, for good or ill. Some of the more unfortunate effects have been mentioned. A more positive aspect is that when we learned to speak we were able to do so more effortlessly and to produce a wider range of sounds.

The question of why one primate, and only one, has acquired the power of speech is another of the unsolved mysteries of human evolution. One approach to the question has consisted of attempts to find out why chimpanzees — although so closely related to ourselves — do not speak.

One early experiment to investigate the chimp's powers of verbal communication, set up by the American psychologists K. J. and Caroline Hayes, met with limited success. They tried to teach their chimpanzee Vicky to talk, and after six years of hard work the animal had learned to utter only four words — 'papa', 'mama', 'cup' and 'up'. Later, when Allen and Beatrice Garner set out to teach sign language to the chimp Washoe, the results were much more impressive. Washoe acquired a vocabulary of 34 signs within two years.

These results were not too surprising. Chimpanzees have an intricate social system requiring subtle methods of communication which is conducted chiefly by means of body language. Postures and facial expressions signal intentions and emotions by eloquently conveying threat, anger, fear, and a wide range of other reactions. Washoe was being asked to extend the range of a channel of communication — that is, sign language — already highly developed and in constant use. Vicky was being asked to renounce it in favour of elaborating the vocal channel of communication. Apes do have a vocal repertoire, but it is far more stereotyped and less flexible than their use of visual signals.

To many people, however, the different responses of Vicky and Washoe meant that (a) chimpanzees have the intelligence to understand the concepts behind such words as 'table', 'hug', 'food', 'open', 'bad', 'into', 'give', and many others; (b) they have the desire to communicate; (c) therefore there must be some specific obstacle which prevents them from speaking the words.

It was implied, and sometimes stated, that chimps cannot speak because the structure of their mouths and throats debars them from producing the requisite sounds. This is not the case. Apes and monkeys can produce a variety of vowel sounds, varying both in pitch and volume. They can say 'ah' and 'ee' and 'oo'. They can pronounce 'k' and 'p' and 'h' and 'm' and the glottal stop (the sound which in Cockney dialect replaces the 'tt' in 'bottle'). They might have difficulty with 't', 'f', and 'n', but there is no structural reason why they could not produce 'b' and 'g'. Anatomists who have studied the question agree that '. . . phonation can be carried on perfectly well by means of an intranarial larynx' — that is, one that has not descended.

The likelihood is that the difficulty Vicky had to overcome was not primarily in her mouth, but lower down in her lungs, where every act of vocal communication starts, and in her brain, because voluntary lung ventilation is a function of the brain.

In most mammals breathing is a function as involuntary as digestion or the beating of the heart. The rapidity of breathing and the amount of air inhaled are controlled by chemical, physical and hormonal factors. For example, a low oxygen level in the blood stream triggers deeper breathing; adrenalin speeds the rate of respiration as well as the heart beat. In higher vertebrates pain, or heat, or cold may alter the respiration rhythm; immersion of the head in water causes cessation of breathing. The receptors which receive these messages are no more within the animals' conscious control than the receptors controlling the knee-jerk reflex. We ourselves are subject to involuntary respiratory reactions of this kind when we sneeze, or sob, or hiccup, or respond to a sudden fright with a sharp intake of breath.

Aquatic mammals, however, have necessarily acquired more conscious control over the operation of their lungs. Unlike land mammals, they can voluntarily regulate the timing of the breaths they take and the amount of air they inhale. Land mammals' breathing reflexes respond only to events which have already happened, as in the case of an animal which accidentally falls into the water. A diving mammal like a seal or a dolphin purposefully regulates its breathing in relation to actions it intends to perform, as explained in a report by R. Elsner and B. Gooden on diving asphyxia:

> Some major component of the diving response is determined by the intention, conditioning, or psychological perspective of the animal being studied. Thus the diver acts as though it produces the most intense diving response when the need for achieving maximum diving duration is anticipated.

Homo sapiens, unlike apes and most land mammals, acquires conscious control of breathing at an early age. A young child, when it is learning to know its own body, discovers that it has power to control its breathing, just as it learns that it has power to control the sphincter regulating its bowel movements. At this stage it enjoys practising and exercising these powers in ways often disconcerting to those around it. If it deliberately refuses to breathe, as some children do, it can hold its breath long enough to terrify them. Some children learn to exploit the trick to express resentment or to practise blackmail.

Adult humans find it easy and natural to respond to a command such as: 'Breathe in — hold it for a count of five — breathe out slowly.' They find it so easy that they seldom appreciate that only a small minority of mammals are capable of this feat. When Vicky was urged to speak she was being asked in effect to perform an act of yoga — to attain conscious control over a function not normally subject to it, as humans do when they are taught to lower their blood pressure by an act of will. Such tricks can be performed, but they are not easy.

The really indispensable pre-adaptation for speech is the enhanced degree of conscious breath control which we share with all diving mammals and no purely terrestrial ones. The pattern of inhaling deeply and quickly, and exhaling slowly at a controlled rate, is characteristic of aquatic mammals when they dive — and of humans when they speak.

12

Sex in Transition

'I never saw an oestrus fossil.'
Tim White

In the field of sex and reproduction, man differs from the African apes in a wide variety of ways. Among the most basic and far-reaching is the absence of oestrus.

In virtually all mammals the oestrus cycle is a vital and efficient system of regulating sexual activity by causing the females to become receptive at certain fixed intervals. It conduces to species survival by ensuring that the greatest procreative efforts take place at the time when they are most likely to produce results, namely, around the time of ovulation. At other times sexual activity is, for the purposes of natural selection, a waste of time and energy and semen.

Oestrus provides a clear signal — olfactory, visual or behavioural — of the time when the female will welcome and co-operate with sexual initiatives by the male. In many species the signal is the trigger of desire in the male. It thus promotes harmony between the sexes by maximising the chances that all sexual encounters are reciprocal, amicable and mutually rewarding.

Since oestrus does not normally occur during pregnancy and lactation, it ensures that the female can concentrate on her maternal role at the proper time without being distracted either by her own desires or by male importunity. Among larger animals native to temperate zones, the cycle is an annual one, and sexual activity is so timed that all the young are born at a time of the year when food will be plentiful. In regions where seasonal variations in the food supply are less extreme, the cycle can be shorter; within a single breeding population, the timing of oestrus and childbirth in different females is staggered. In the African apes the average length of the cycle is 31–32 days in the gorillas and 35–36 days in the chimpanzee, as compared to 29 days in humans.

The oestrus system, in short (like fur and quadrupedalism), is a device which has evolved over millions of years, is almost universal among mammals, has multiple advantages and works very successfully.

These advantages have been lost to our species since oestrus is not a feature of the human sexual cycle. In most other respects the female cycle in humans conforms to the anthropoid norm. As in apes, menstruation marks the end of the cycle, and ovulation occurs roughly halfway through it. But in humans alone ovulation is unaccompanied by any peak of desire in the female, and is not detectable by the male.

From time to time attempts are made to prove that oestrus is still operative in humans, but the evidence offered is usually anecdotal rather than scientific. One survey, based on a questionnaire distributed among students, reported a tendency for sexual activity in females to be slightly more frequent somewhere around the time of ovulation rather than during other parts of the cycle. But this fact is largely accounted for by avoidance of sexual activity during and immediately prior to menstruation. In any case, the tendency if it exists at all is so vestigial as to be of no practical value. This peculiarly human phenomenon is known to scientists as 'concealed ovulation'.

As with other anomalous human features, there is a

tendency to interpret the suppression of oestrus not as the loss of a valuable heritage, but as a kind of evolutionary Great Leap Forward which only one species is sufficiently advanced to have achieved. Oestrus is sometimes thought of as one of the more 'bestial' aspects of animal sex, and associated with promiscuous couplings. But it is equally adaptive in strictly monogamous species, such as the wolf and the gibbon, for a female's peaks of desire to coincide with her peaks of attractiveness and fertility. It is hard to think of anything better designed to cement the pair bond.

The need to prove the superiority of an anoestrous species was hampered by the fact that no one was quite sure what had happened to the formerly periodic female libido. One theory holds that it has vanished. Another suggests that human females are not merely 'permanently receptive' but in effect permanently on heat. Both versions are represented as advantageous for the species.

The first version proposes that oestrus became counterproductive during the hunting/gathering stage of human evolution because the presence of sexy females interfered with male bonding. The males had to develop a spirit of comradeship in order to co-operate on the hunt for game, and '. . . consequently they could no longer afford socially disruptive aphrodisiacs. Among the males, there would be occasion enough for tension — disputes over distribution of meat, quarrels over desirable mates — without having the tension increased by the sight and scent of oestrous females.'

The second version suggests that food supplies on the savannah were scarcer and less dependable than in the forest; that rearing the young successfully demanded active co-operation from the males, best assured by a system of pair-bonding; and that the best way of cementing the pair-bond and inducing the males to contribute to the food supply was for the females to reward them by becoming 'permanently receptive' and 'making sex sexier'.

The theory won widespread acclaim, and it is often retailed in the textbooks as representing the conventional wisdom. It was not embraced because of any inherent probability. It was welcomed because if you start with the axiom that the loss of oestrus *must* be interpreted as a gain rather than a loss, then you are reduced to grabbing at straws. The theory depends on a chain of implicit assumptions. Not one of them is proven, and some of them are absurd.

The theory implies, firstly, that being 'permanently receptive' was a new departure in hominid females, not found in other apes. If 'receptive' means merely permitting copulation rather than desiring or soliciting it, this is not true. A young female chimpanzee not in oestrus will frequently defuse the anger of an aggressive mate by presenting her rump and allowing him to mount her, though the mounting does not always lead to intromission. The interaction has more to do with appeasement than with sexual bonding (a young male will perform exactly the same gesture when threatened with attack), but in this sense female chimpanzees are receptive most of the time.

Secondly, it implies that male co-operation in the care of the young must be secured by a system of rewards such as unlimited sexual access. This is a fallacy. Male Emperor penguins do not have to be bribed to huddle fasting on the ice, incubating eggs through the blizzards of the long, dark Antarctic winter. Where paternal instinct operates, it is a categorical imperative in the same way as maternal instinct. There is no known species in which it has to be reinforced by sweeteners of any kind.

Thirdly, it implies that in pair-bonding primates the couple are held together by frequent passionate sexual activity. This is the reverse of the truth. The only monogamous apes are the gibbons. The male gibbon has a hundred per cent certainty that any offspring born to his mate will carry his own genes, and consequently does not need to over-exert himself. Sexual activity is at a lower level in gibbons than in other ape species.

Fourthly, it implies that 'permanent receptivity' would automatically promote pair-bonding — in other words, that living in a society where all the females were on heat all of the time would ensure that each male remained faithful to one of them for life. It sounds highly improbable.

Finally, it makes the large assumption that the hominids were pair-bonded. This claim is frequently made. It is a central plank in some of the conventional scenarios of foraging males and stay-at-home females. But it remains unproven.

It is quite true, as implied by Tim White's terse comment at the head of this chapter, that fossilised bones do not tell us very much about the sexual behaviour of our prehistoric ancestors. On the other hand, it is often possible for an anatomist, by studying the physiology of a species, to learn a great deal about its social organisation even before its behaviour has been studied by ethologists in the wild.

For example, one common kind of social grouping is described as 'harem-type'. It consists of one dominant male accompanied by a number of females. Because the male is monopolising more than his share of the females, his position is frequently challenged by rival males and he must be equipped to defend it. The typical hallmarks of harem-type societies are that the male is much larger than the female and is equipped with natural weapons such as tusks, horns or antlers.

The gorilla conforms to this pattern. The male is twice as heavy as the female and has long, strong canine teeth, which are not found in the females. Man is obviously not cast in this mould. The size difference between men and women is very much less and the natural weapons are absent. A man's canine teeth are relatively no larger than a woman's.

Chimpanzees live in larger and more anarchic groups. Occasionally a male will establish a short-term monopoly of one particular female, but most of the sexual activity in a troop of chimps is opportunistic — and opportunities are frequent. Not only do females tend to outnumber males in a chimp community, but each female is in oestrus for ten days of every cycle as compared with two days in the gorilla, and during that period she may be mated by most or all of the males.

The most effective male strategy in such a promiscuous society is frequent hasty copulation, which demands an abundance of sperm. So the hallmarks of that kind of organisation are, firstly, the retention of natural weapons — because the system gives rise to a good deal of male rivalry and aggression — and secondly, the possession of very large testes. Although a chimpanzee is less than a third of the size of a gorilla, his testes are four times as large. They are also four times the size of a man's.

Gibbons are monogamous, and they display the hallmarks of monogamous species in that there is virtually no difference between the sexes either in body size or in the length of the canine teeth.

In physical terms man conforms to none of these patterns. The size difference between the sexes (20 per cent on average, as compared with 100 per cent in the gorilla) is not great enough to indicate a harem-type society, but is too marked to suggest a monogamous one. Nor does the size of the human testes indicate monogamy. While they cannot compare with the chimpanzee's, they are appreciably larger than the gorilla's, and a man can attain an ejaculation frequency of 3.5 times a week before his sperm count begins to drop. This suggests a society which allows for a fair amount of opportunistic sexual activity. The large comparative size of the penis in adult male humans (man 13 cm; chimpanzee 8cm; gorilla 3cm) is not related to frequency of deployment. It is a necessary corollary of the retraction and relative inaccessibility of the vagina.

An aquatic environment seems to have had a broadly similar effect on some other species — that is, relative retraction of the female sex organ leading to a corresponding extension of that of the males. For example, most birds and reptiles do not possess a penis; the pressing together of the cloacal apertures seems to suffice for the transference of the sperm. But many species of aquatic reptiles (crocodiles and turtles) and aquatic birds (swans, ducks, geese) have found it necessary to evolve a penis as part of their adaptation to a watery habitat.

Overall, the anatomical evidence in respect of *Homo* suggests a species in a state of transition. There is nothing very surprising in this conclusion, for a similar transition must have taken place also among the great apes. The chimp/gorilla split occurred almost as recently as the chimp/human split, yet one or both of the African apes must have changed its sexual strategy quite radically since then, or they would not behave so differently today.

The transition, in the case of the hominids, appears to have been in the direction of a more monogamous society. Humans are more prone to form ongoing one-to-one sexual relationships than chimpanzees are, and their consortships, once formed, tend to last longer. A trend towards pair-bonding would be an entirely natural development in a species where the business of rearing the young became increasingly onerous as the period of infantile dependency grew longer.

It is true that monogamy does not come to us quite as naturally as Western cultures have often implied, and that of societies untouched by Western teaching only 26 per cent are monogamous. The existence of adultery and the high divorce rates are evidence that our instinct for pair-bonding is by no means absolute. On the other hand, the majority of divorcees re-marry; serial monogamy is still monogamy. And even among pair-bonded species such as gibbons, each of the pair is occasionally tempted to stray unless the would-be interloping males and females are chased away by the aggrieved partner. It would seem that the price of pair-bonding in gibbon society, like the price of freedom, is constant vigilance.

It is tempting to blame all sexual tensions and difficulties on 'society' — on more and more taboos — on inculcated feelings of shame and guilt, economic pressures, religious dogmas and artificial gender stereotypes. All these undoubtedly add to the difficulties. But it would be a mistake to imagine that there was a time, prior to civilisation, when Nature had arrived at a perfect solution.

All the orthodox theories on oestrus suffer from the same weakness as the other savannah scenarios, namely, that numerous other mammal species have faced precisely the same problems but none has resorted to the same solution. She-wolves, for example, depend on male co-operation in raising their young. The strategy adopted by wolves exactly mirrors the popular foraging-male scenarios proposed for the hominids. There is a female staying at home in a lair with the cubs; there is a male hunting far afield and bringing back food to his family. There is lifelong pair-bonding. Yet the female retains oestrus, because in a pair-bonded species its value is, if anything, enhanced. It continues to promote harmony between the sexes, without any danger of causing friction between the males, since each of them is interested only in its own mate.

When we are confronted with a unique characteristic like the suppression of oestrus, it is surely more probable that it was a response to a unique predicament, rather than an anomalous reaction to a very common one.

Th predicament in this case could have been an aquatic habitat. In mammals, oestrous status is communicated by scent signalling — a pheromonal message emitted by the female. Being airborne, it may be carried quite a long way — as evidenced by the distance a dog will travel to locate a bitch on heat. But in a wading or swimming ape the pheromones would be washed away almost as soon as they were secreted.

Some anthropologists are inclined to dispute this proposition by querying the premise: that oestrus is always primarily a scent signal. This is because in some anthropoids there are also visual signs of oestrus. In chimpanzees, for example, the skin around the vagina periodically swells up and becomes pink and conspicuous, reaching a peak of tumescence during oestrus. A similar phenomenon occurs in baboons and macaques. It has been interpreted as a visual signal directed at the male.

It is unlikely that it originally evolved for that purpose. Sexual swelling is also found in the gorilla, but since it is less extensive and the skin is covered with hair, its value as a visual signal is nil. The degree of sexual swelling, as between species, does not correlate with anything in the sex life of the animals concerned. It correlates with the habitat. With the exception of the talapoin monkey and forest-dwelling chimpanzees, the species concerned are all ground dwellers, and the degree of sexual swelling appears to relate to the extent to which the monkeys or apes spend their time squatting on hard ground. It is most extensive in species living on the open savannah or in rocky habitats (for example, baboons and geladas) and least extensive in species dwelling in the forests (such as gorillas and mandrills) where the terrain is softer and leafier. It could have evolved originally as a protection for the female, to minimise direct contact between the ground and the vaginal opening during oestrus, when the female's own secretions plus traces of semen could cause dust and grit to adhere and be driven in during intercourse.

Ethologists, being human, have a very limited sense of smell. To them the visual signal conveyed by the pink backside is the most conspicuous sign of oestrus in the chimpanzee; it has been an invaluable aid to them in relating the animals' behaviour to the stages of their sexual cycle. It must be tempting for them to assume that it is also the most obvious and important signal to a male chimpanzee. But experiments have shown that the male is responding primarily to a pheromonal signal. A male chimp's response to an oestrous female is unaffected when the visual signal is concealed or disguised, whereas he rapidly loses interest if the olfactory signal is counteracted.

On balance there seems no good reason to doubt that in oestrous mammals the signal perceived by the male is primarily olfactory. But in humans the ability to receive and interpret scent signals is very low. The olfactory lobe in our brains is proportionately smaller than in the brains of apes. (This is a common feature in aquatic mammals. In whales and seals the olfactory lobe has diminished almost to vanishing point.) So one reason for the ending of oestrus could be that it ceased to work properly. As a result of the pheromonal secretions being washed away, plus diminished scent perception, the signal was simply not getting across.

Another possibility is that it ceased to matter whether this signal got across or not, because male behaviour was no longer being regulated by the rhythms of the female cycle. To understand how this can happen, it is necessary to consider a major landmark in human sexual evolution — the change to face-to-face (ventro-ventral) copulation, the commonest mode in humans. This was a direct result of the stretched posture, with spine and hind limbs in a straight line, which results from either swimming or bipedalism.

In most quadrupedal mammals the vagina is accessibly located at the rear, with the opening more or less flush with the surface of the body and covered only by the tail (if there is one — apes have no tails.) In *Homo sapiens*, as a result of the physical modification for upright walking, the vagina is directed ventrally — towards the front. It is also retracted inside the body wall and covered by thick folds of skin (the *labia majora*) which are not found in the apes.

Bipedalism alone does not account for all the changes. It is rare in land animals, but quite common among aquatic ones, for the vagina to be retracted and covered with a flap of skin. The same is true of another human feature — the hymen. The hymen is commonly found in small primitive animals, and its original function was probably to ensure that sexual activity was limited to the period when conception was most likely. In the guinea pig, for example, the membrane reseals the vagina after each reproductive period. The hymen is still found in the lemurs of Madagascar and other prosimians, but not in monkeys.

It seems unlikely that its presence in *Homo* can be accounted for by thinking of it as a primitive hangover. An aquatic connection seems the likeliest explanation. K. E. Fichtelius has pointed out that a well developed hymen is also a feature of various aquatic mammals quite unrelated to one another, such as the toothed whales, and the seals, and the dugongs.

Ventro-ventral copulation, very rare in land mammals, is the commonest mode in aquatic mammals except for those which go ashore to breed. Whales and dolphins, dugongs and manatees, beavers, and sea otters are among the numerous aquatic species which mate face to face. Swimming promotes this method of copulation in the same way that bipedalism does, because in both cases the spine and the hind limbs are realigned, forming a continuous straight line instead of the 90-degree angle found in most quadrupeds.

In the early '70s the suggestion was made that for a land mammal, changing to the ventro-ventral position would initially have proved traumatic for the females. In all quadrupeds the supine position is a particularly vulnerable one. Among primates there are fairly strict rules governing the infringement of personal space. Close encounters among adults for benign purposes — copulation or grooming — are normally conducted by an approach from the rear. A purposive approach by an adult male from the front usually signifies aggressive intentions. And primates — unlike many birds — have no repertoire of courtship ritual or foreplay which might give assurance to the female.

This suggestion was discounted as pure speculation. It could only be substantiated by finding another anthropoid species in which ventro-ventral copulation was the commonest mode, and observing how it affected the relationship between the sexes. No such species was known to exist. Since then studies have been made of the sexual behaviour of orang-utans and have resulted in some remarkable findings.

In a study by R. D. Nadler in 1977, four pairs of orang-utans were kept under observation. The male of each pair was allowed daily access to the females. Three of the four pairs mated every day throughout the female's sexual cycle, sometimes more than once. (The female orang-utan is therefore nowadays technically classified, like the female human, as 'permanently receptive'.) Copulation was ventro-ventral.

> Essentially all copulations, moreover, were initiated by the male in a manner resembling rape . . . At the beginning of each test, the male pursued the female, who fled and frequently sought to escape by climbing to the top of the cage. The male quickly caught the female, wrestling her, struggling and screaming, to the ground and forcibly initiated copulation. Once the male achieved intromission and initiated thrusting, the female became passive.

However, primate behaviour in zoo and laboratory conditions is often atypical, so it was important to confirm these findings by studies of behaviour in the wild. B.M.F. Galdikas observed wild orang-utans in Borneo over a period of four years from 1971–1975. Copulation was less frequent than under laboratory conditions but, in the wild as in captivity, it was 'generally ventro-ventral' and frequently carried out by force. The females' struggles '. . . ranged in intensity and duration all the way from brief tussles with squealing and some pushing and slapping at the male's hands to protracted violent fights in which the female struggled throughout the length of the copulation, emitted loud rape grunts, and bit the male whenever she could.'

The frontal approach in orang-utans is not attributable either to water or to bipedalism, but to a combination of size and habitat. The Asian apes — the gibbon and the orang-utan — are far more arboreal than the African ones. The gibbon is the smallest and most agile of all the apes and swings through the branches as lithely as any monkey, but the orang-utan is much larger and slower. It cannot afford to launch itself into space because of the risk of landing on a branch that would not bear its weight. The gorilla, when it grew large enough to be confronted by the same problem, came down to the ground where it now spends virtually all its waking hours. That option was not open to the orang-utan because it lives on fruit rather than greenery. It stayed aloft to be near the food supply and it lives and moves ponderously, suspended from the branches by its hands and often by one or both of its prehensile feet as well.

For such a large animal, sex in the tree tops presents problems. The male lives alone but his territory overlaps with that of around four females who constitute his 'dispersed harem'. Mating in the normal quadrupedal position is impossible because a female could never keep her balance on top of a branch while sustaining the weight of a male twice her size. Copulation from the rear with both animals in suspension would also be difficult: the male would have at best one hand free to grasp the female and intromission would be hard to effect with her body swinging free in the air like a pendulum. In practice, he corners her in a suitable place where she can hold on to one branch and recline on another, and forcibly manoeuvres her into a supine position. It creates a good deal of commotion, but that is of no significance because the animals are not threatened by any predators other than man.

There is no way of establishing just how recently these animals made the transition to ventro-ventral mating as a standard procedure. It is almost certainly much more recent than in the case of our own species. The reasons for thinking this are anatomical and behavioural. The orang-utan's sexual organs are less modified than in humans to accommodate the ventro-ventral approach, and the females' behaviour suggests either that the males' approach is inept or even that they do not at first understand what is happening. In the wild, as in the laboratory, the fighting usually — though not always — ceases once the male achieves intromission and the female realises that he is bent on sex and not on mayhem. From that point on resistance is often succeeded by restive resignation — ('fidgeted and looked around') – or a detached passivity — ('resumed eating, reaching out and plucking fruits as the male was thrusting'). On some occasions the mating is wholly co-operative with no fighting, when the female is mature, and experienced, and is in oestrus.

For the orang-utan has not yet entirely lost oestrus. It is not easy to detect because, being arboreal, the animals display no visual signs of it in the shape of sexual swellings. However, its presence was demonstrated in a neatly designed study. Male and female orang-utans were placed in two separate compartments of a test cage, and the doorway between the compartments was of such a size that the female could pass through it at will but the male, being larger, could not go through it at all. In previous experiments where the male had free access, copulation took place every day. When the female was able to regulate access, it took place only during a brief period in the middle of her 29-day cycle — that is, around the time of ovulation.

This clearly indicates that the anthropological conception of being 'permanently receptive' has very little to do with the female and her own inclinations. In the case of orang-utans the male ignores the presence or absence of the oestrous signal because it no longer serves one of its primary purposes — that of assuring him of a welcoming and co-operative response to his advances. The chances are that he will have a fight on his hands at any time of the month, especially in the case of younger females, whenever he approaches them. Once that situation arose, the males which left most offspring would have been those which were least inhibited from using force to effect copulation, and that has become characteristic of the species. In such circumstances, frequency of copulation does not indicate that the female is permanently receptive, but merely that the male is permanently appetitive and not easily deterred. For all practical purposes, the oestrus system has broken down, and in course of time may eventually cease to exist.

In humans it has already ceased to exist. Desire in women is no longer triggered from the inside by cyclical change in their hormonal balance. In biological terms it became irrelevant to species survival: male desire was sufficiently active and dominant to ensure that females reproduced their kind whether or not they invited or welcomed the process of procreation. When any biological feature becomes irrelevant to survival, the forces of natural selection cease to exercise any influence over it, whether in the direction of changing it or stabilising it. One result of this may be a wide degree of variation in the feature as between individuals. An example is the vermiform appendix — a hollow blind-ended tube attached to the large intestine. In man, since it has ceased to perform its original function (the digestion of cellulose), it has become vestigial and relatively unpredictable in form. Its average length is between three and four inches, but it may in some cases be as long as seven inches, and in others virtually non-existent.

It is highly likely that the strength of sexual desire in women is subject to innate variation. But this is hard to quantify because of the powerful influence of social conditioning. The conditioning factors include early upbringing, education, social and cultural backgrounds, peer group pressures and life events, especially those experienced during childhood and puberty. The same factors affect males but to a lesser degree because the instinctive component in their sex behaviour is less eroded.

Women discussing their experiences of love and sex are liable to encounter barriers of incomprehension when talking to women from very different cultures — or even, in an era of rapid social change, to women of different generations. There is a prevalent idea that somewhere in the world there must be, or have been, a subsection of womankind which got it *right*, exercising simple natural sexual responses, able to illustrate the 'real truth' about what makes all women tick. If that were true, the corollary would be that all other women before them — or after them, or living elsewhere, or dissenting from them — have always been self-deluded, or brain washed, or telling lies.

That is a fallacy. It is like believing that somewhere in the world is a tribe or race speaking the language that Nature intended people to speak, from which all the rest of us have deviated. But it is a fallacy widely believed in, and women are under pressure to conform to the current stereotype for fear of being castigated in one century for unnatural depravity, in another for unnatural frigidity.

To return to the question of frontal sex, the signs are that in *Homo* the transition to this mode happened much earlier than in the orang-utan, so that we have had several million years in which to adjust to it, both behaviourally and physically. We have evolved subtler methods, verbal and non-verbal, of communicating our intentions to one another. And in the course of evolution the human vagina has swung forward to accommodate ventro-ventral copulation, so that its axis now forms an angle of more than 90 degrees with that of the uterus instead of being in a straight line with it, as in most quadrupeds.

If, as this suggests, the change occurred at an early stage of hominid speciation — possibly before Lucy — it is conceivable that when it first took place the females' responses could have been as disturbed and resistant as the orang-utan's, and equally disruptive of sexual harmony. Perhaps we should count among the scars of our evolution the atavistic psychological link between sex and violence which appears to be a human characteristic but is certainly not part of the general mammalian inheritance.

In most wild animals sex and hostility are incompatible. Sexual activity may give rise to hostility between males, but not between the partners in the sex act. In the majority of quadrupeds rape is unknown. Not only is there no incentive for it (because of oestrus), but it is a physical impossibility. If a mare or a vixen or a she cat does not want to be mated, she has only to hold her tail down and keep running — or even keep walking — and it cannot happen. If she is receptive, she stands still. We belong to the small minority of animal species in which rape is possible, and to the still smaller minority in which it occasionally takes place.

Some people find it distasteful even to think of these matters in terms of our own animal origins. Humans are endowed with intelligence, and foresight, and sensitivity. We are not at the mercy of our instincts. Loving relationships between humans are of infinitely greater complexity than relationships between animals.

But the fact that they are not purely instinctive

behaviour patterns is no guarantee of a foolproof system — rather the contrary. Instinctive behaviour can exhibit a remarkable degree of efficiency and reliability. The near-perfection of a swallow's navigation, a spider's web, or a weaver bird's nest is achieved precisely *because* these functions depend on an innate capacity and do not have to be learned. In lower species the same is true of sex.

But in many primates, including man, sex is a learned activity, and not one for which they are fully programmed at birth. If monkeys are reared in isolation from their own kind up to the age of two years and then brought into contact with one another, they never discover by themselves through trial and error how to copulate. Some of the female isolates may finally mate successfully if partnered by a proficient male, but the male isolates never achieve intromission.

Much the same is true of *Homo sapiens*. In Victorian days there were rare cases (documented in medical records) of young men strictly reared who grew to adulthood without ever having been told the facts of life, and when they married young women equally ill-informed the union remained unconsummated. In a species where so powerful an urge is accompanied by so little instinctive knowledge of what to do about it, there is a tendency for the elders of the tribe to issue their own list of instructions and proclaim that any who depart from it are accursed.

People who try to prescribe or reform sexual mores often claim that the behaviour they advocate is what Nature intended, as if that were in itself a sufficient recommendation. It is not really a very good criterion. 'Nature' — as the personification of our own evolutionary inheritance — is not all-wise. She proceeds by devising *ad hoc* solutions to environmental crises and vicissitudes, and they are often remarkably successful. But in the sphere of sexual relationships she has endowed us with a confused and messy legacy.

The human problem is how to make the best of it in the rapidly changing modern world. Perhaps it is time to ignore where we have come from and concentrate on where we want to go. After all, the peculiar weaknesses of our species are more than counter-balanced by its peculiar strengths which are still under-exploited — the ability to think, to communicate, and to foresee the long-term consequences of our decisions on our own well-being and that of our successors.

13

The Aquatic Ape Theory — the Counterarguments

'Often the older scientists of the discipline
are unable to make the switch; they feel
strong hostility to the new paradigm and its
supporters; and matters are only resolved as
the old-timers die off (Plank's Principle).'

Michael Ruse

Plank's Principle means that a whole generation frequently elapses before the scientific community is prepared to consider a new hypothesis dispassionately, on
its merits. In that intervening period the reception
accorded to the unfamiliar idea has little to do with merit.
Stephen Jay Gould has described the early reactions to
Wegener's theory of Continental Drift (see p.50)

During the period of nearly universal rejection, direct
evidence for continental drift — that is, the data gathered from rocks exposed on our continents — was
every bit as good as it is today. It was dismissed
because no one had devised a physical mechanism
that would permit continents to plow through an
apparently solid oceanic floor . . . Since drift seemed
absurd in the absence of a mechanism, orthodox geologists set out to render the impressive evidence for
it as a series of unconnected coincidences.

That is happening again today in the life sciences. There is an impressive list of unusual features in the human body which need to be accounted for — nakedness, bipedalism, fat, tears, the sebaceous glands, the descended larynx, the vanished apocrines, and so on. The Aquatic Ape Theory (AAT) suggests that a single factor may account for the great majority of them. Orthodox theories leave a number of them unaccounted for, and for the rest they persist in attributing the multiplicity of unique features to a multiplicity of possible causes, as a '. . . series of unconnected coincidences'.

A theory which obviates the need to postulate unconnected coincidences is said to possess the scientific virtue of parsimony. It was this aspect of his own theory which kept Darwin going when his ideas met with scepticism. He wrote: 'I must freely confess, the difficulties and objections are terrific; but I cannot believe that a false theory would explain, as it seems to me it does, so many classes of facts.'

One kind of response to an unwelcome idea is to ignore it, and this can be very effective. If the new concept has nothing going for it except the vividness of the creator's imagination, it will sink into oblivion. This policy was initially applied to AAT and appeared at first to be successful.

In 1930, when Alister Hardy as a young marine biologist conceived the idea of an aquatic phase in human evolution, his friends warned him that publishing such a heresy would mean committing professional suicide. As he admitted in later years, 'I wanted a good professorship. I wanted to be a member of the Royal Society.' So he kept quiet about it for the next thirty years. When he finally published his idea in 1960 the response from the scientific community was nil.

Meanwhile, the same possibility had been considered in Europe. It is not unusual, in the case of an idea whose time has come, for it to occur independently to more than one person. (The most famous example is that of Natural Selection, which was developed independently by Charles Darwin and the naturalist Alfred Russel Wallace.) A German professor, Max Westenhöfer, wrote a book on human evolution in 1942, which included a chapter on the aquatic hypothesis. It was ignored so successfully that few people outside Germany learned about it, and Hardy died without ever knowing that he had not been the first to publish the idea.

Nevertheless, the aquatic hypothesis refused to go away. Despite the chilly reception accorded to it, it survived and gained adherents. The time came when tutors in Anthropological seminars began to complain of a recurring nuisance — the kind of enquiring student who would stand up and ask questions about AAT.

The strongest response would have been to say: 'We can explain all the anomalies more elegantly and convincingly without needing to postulate an aquatic detour, and here are the explanations.' Instead of that, students were liable to be told: 'The explanations are all there in the literature — go away and look them up.' One object of writing this book has been to come to the aid of such students by saving them a lot of time and trouble. To the best of my knowledge, all the non-aquatic explanations to be found in the literature have been indicated in the previous chapters.

Another response is a refusal to discuss the matter on high philosophical grounds. AAT, it is claimed, does not merit attention because it is not, in the true sense, scientific. It makes no 'verifiable predictions'; it 'can never be falsified'; it changes its arguments in the light of new discoveries; the convictions of its adherents 'cannot be shaken by contrary evidence'.

This is an intimidating catalogue, indeed — until you apply the same criteria to the scientific credentials of the savannah theory. It, too, has made no verifiable predictions; it, too, alters its scenario with every new discovery.

The savannah carnivore becomes a savannah seed-eater; the hominid slowly evolving on the plain becomes a hominid evolving there with spectacular speed; the large-brained tool-user gradually turning into a bipedalist is transformed into a weaponless biped gradually acquiring a large brain.

Through all these vicissitudes nothing shakes the conviction of the savannah theory supporters that basically they have been right all along and are still right. There is no basis whatever for the claim that the savannah theory is in some mystic sense sounder or 'more scientific' than AAT.

For some hard-working anthropologists, Hardy's idea is disqualified from being taken seriously by the manner in which he arrived at it, and by the enthusiasm of some who accept his ideas. He came home from an Antarctic expedition, opened a book by Frederick Wood Jones, and almost at once all the things he knew about man and evolution rearranged themselves into a new pattern in his head.

His critics feel that in this day and age scientists do not and should not receive enlightenment in this dramatic fashion, like St Paul on the road to Damascus. New truths, they maintain, should be vouchsafed only to those who have earned them by years of painstaking endeavour — and by themselves providing some of the new data paving the way to the new theory. After all, it is urged, we are not living in the age of Archimedes, and the man Hardy was not even an anthropologist. He was a marine biologist, and should have stuck to his own speciality.

But the 'Eureka' experience did not go out of fashion with Archimedes. The idea of natural selection occurred to Alfred Russel Wallace during a bout of malaria in the East Indies in 1858, and the impact of it shot him out of his sickbed in search of pen and paper and a candle. When he related this to Charles Darwin, Darwin pointed out to him the precise spot where he himself, when riding in his carriage, had experienced the same excitement.

Michael Ruse, in *The Darwinian Paradigm*, relates what happened in the present century when Wegener's half-forgotten ideas resurfaced. 'We have seen already how liberating and exciting geologists find their new theory, and it seems clear that many of them come to the theory by something very much akin to a conversion experience.' One researcher, Tanya Atwater, who was in at the beginning of the geological revolution, recalls:

> It is a wondrous thing to have the random facts in one's head suddenly fall into the slots of an entirely orderly framework. It is like an explosion inside . . . I took my ideas to John Crowell on Thanksgiving Day. I crept in feeling very self-conscious and embarrassed that I was trying to tell him about land geology starting from ocean geology, using paper and scissors. He was very patient with my long bumbling, but near the end he got terribly excited and I could feel the explosion in his head. He suddenly stopped me and rushed into the other room to show me a map of when and where he had evidence of activity on the San Andreas system. The predicted pattern was all there. We just stood and stared.

There may be scientists so austere that they would merely have lifted an eyebrow. But it is clearly untenable to imply that any idea which generates the sensation of an explosion in the head must *ipso facto* be invalid.

More recently some scientists have abandoned these abstract admonitions and begun to address themselves to the evidence. Sometimes they repeat the error of bringing charges against AAT which could equally be levelled against orthodox theory.

For example, it has been asserted that a naked skin cannot be regarded as an aquatic adaptation because not all aquatic mammals have lost their hair. By that same reasoning it certainly cannot be regarded as a savannah adaptation.

Similarly, the argument that 'man's ancestors can never have been aquatic because some people fear water' is no more valid than 'man's ancestors could never have been arboreal because some people are afraid of heights'. The more substantial arguments are advanced by those who are willing to try to analyse their instinctive rejection of AAT, and to give reasons why they think the aquatic interlude could never have taken place.

One such contention is that an aquatic ape could never have survived because it would have died of cold. Primates, the argument runs, are homeotherms and have to maintain a minimal blood temperature or perish; in water they would be in danger because water conducts heat away from the body more rapidly than air does. Data are produced purporting to prove that in water at a temperature below 28°C the body's core temperature will begin to decline, and that any figure below 23°C entails imminent danger of death from cardiac arrest.

This sounds like a cogent argument but, like the theoretical proof that it is impossible for a bee to become airborne, it does not work in practice. If it did, then no warm-blooded species could ever have adapted to life in water, and a large number of mammals have, in fact, done so.

It does not hold true of modern humans, either. Diving women off the shores of Korea and southern Japan earn their living by diving for shellfish and edible seaweed. In summer they work for four hours a day wearing loin cloths, sometimes descending to depths of 80 feet. In winter, when the water temperature sinks to 10°C they work for shorter periods in light cotton bathing suits. Kirsten Schuitema, a zoologist from the University of Lund, reports that '. . . adults in one group in Indonesia spent a daily average of 6 hours in water, and the children

4–5 hours.' Individuals spent up to ten hours daily in the sea in periods of 20–90 minutes with short breaks for bringing the catch home.

In 1987 the American Lynn Cox swam from Alaska to the USSR across the Bering Strait, without wet-suit or covering of lanolin. She swam for four hours in water of 7°C maximum, dropping to a low of 3°C — that is, 20 degrees lower than the allegedly fatal figure. She did not die of cardiac arrest. There must be some factor or factors which are being left out of account in the abstract calculations about putative loss of body heat. One obvious factor is the efficiency of the fat layer as an insulator in water, but that is clearly not the whole story.

Donald W. Rennie conducted an experiment with diving women lying in a tank of water for three hours with only the face above the surface. Other Korean women from the same community — who were not divers but had the same thickness of subcutaneous fat — were immersed in the same way. Rennie found that the divers suffered less heat loss than non-divers with an equivalent fat layer. He, and other researchers in the same field, have deduced that there must be some latent human mechanism of adaptation to cold stress. Rennie suggests that '. . . the divers' fatty insulation is supplemented by some kind of vascular adaptation that restricts the loss of heat from the blood vessels to the skin, particularly in the arms and legs.'

A mammal living in water would probably need to eat more than a land mammal of the same size. The Korean *amas* (the diving women) consume on average 50 per cent more calories a day than a non-diving Korean woman of comparable age.

But this would hardly have posed a problem to the aquatic ape. Some of the richest food sources in the world are found in tropical wetlands and off-shore waters. The food supply, in quantity, variety and ease of procurement, would far exceed what the savannah could offer to a small unspecialised primate.

The resources available to baboons, for example, consist mainly of leaves, seeds and seed pods, grubbed-up roots and corms, and occasionally berries. Foraging methods may vary — the baboon pulls up grass rhizomes with its teeth, and the gelada tugs them out with its hands — but the savannah menu offers such a meagre and monotonous selection that foraging can be a virtually non-stop occupation. For an early hominid like *Australopithecus* the savannah would have been equally inhospitable. In time they would have learned to supplement their diet by scavenging or opportunist predation on small animals, as some baboons have now learned to do. But as scavengers they would have been in competition with species both fiercer and more specialised, from hyenas to vultures. In the early stages of adaptation to bipedalism they would have been without the speed or the natural or artificial weapons to equip them for the job.

By contrast, coastal food resources include an abundance of edible aquatic plants, fish, crustaceans, bivalves, molluscs and other invertebrates, and the eggs of sea birds, apart from occasional flotsam such as the carcases of dugongs or turtles. Most of these items are available all the year round, and would have enabled an aquatic primate to move from a vegetarian diet to an omnivorous one without waiting to develop the skills and weapons necessary for hunting game on land. For example, some chimpanzees already crack open nuts: an aquatic ape would have had no difficulty cracking open small shell-fish, as crab-eating macaques already do.

A change of diet can have far-reaching effects on a species. In this instance it may have been a factor in the development of the one unique feature we are most inclined to pride ourselves on — the human brain.

14

Brains and Baboons

'The surest and best characteristic of
a well-founded and extensive deduction
is when verification of it springs up, as
it were spontaneously, into notice from
quarters whence they might be least
expected.'

J. F. W. Herschel

At one time it was assumed that the crucial change in
man's cerebral evolution began with *Homo habilis* — that
is, much later than the fossil gap. However, Robert Martin
in 1982 re-examined the evidence from fossils and comparative anatomy. With the use of scaling analysis he
demonstrated that hominid brain growth relative to that
of other anthropoids began at least five million years ago.
It was not a sudden event but a progressive phenomenon,
though it apparently accelerated about the time of *H.
habilis*.

Robert Martin's approach was an original one. Instead
of asking why brain growth might have been adaptive or
desirable, he asked what made it more *attainable* in one
particular primate than in others. In energy terms brain
tissue, as compared with other body tissues, is uniquely
expensive to produce and maintain, making heavy
demands on the mother during pregnancy and lactation.
That may be one reason why there is in many species a
correlation between relative brain size and different diets.

For example, fruit bats' brains are twice the size of the brains of insect-eating bats. Similarly, leaf-eating monkeys have relatively smaller brains than fruit-eating ones of closely related species. A move from a forest diet to a savannah one does not seem to facilitate brain growth. Among cercopithecine monkey species, for example — the family that includes baboons, mandrills and patas and rhesus monkeys — there is no distinction in brain size between forest-dwelling and savannah-dwelling species.

Martin concluded: '*Homo sapiens* must have evolved in response to a rather unusual combination of environmental factors which made available a relatively steady, predictable supply of food lacking in significant toxic levels.'

Michael Crawford in 1989 speculated that a seashore diet could have had special advantages in relation to brain growth in addition to those of abundance, variety and year-round availability. He pointed out that the building of brain tissue needs a consistent one-to-one balance between Omega–6 and Omega–3 fatty acids. These two types of polyunsaturated fatty acids are both essential for health, and the body cannot use one in place of the other. The latter type (Omega–3) are relatively scarce in the land food chain, but predominate in the marine food chain.

The most popular attitude to the question of brain size has always been the idea that having more room inside his skull enabled Man to become more intelligent. This may be described as the Red Riding Hood approach: 'Oh, *Homo sapiens*, what a big head you've got.' 'All the better to think with, my child.'

This is a questionable assumption. Within our own species intelligence bears no relation to the size of the skull — either its absolute size or its relative size. Even in today's complex civilisation we only utilise a small fraction of the brain's potential, and brain scans have shown that people with up to half of the brain destroyed can lead a normal life.

As between species, there is no correlation between

skull size and intelligence. Recent researches into the intelligence of parrots have led some people to re-class them as 'honorary primates'. In the bird's small brain there is an ability to categorise objects and grasp abstract concepts like similarity of colour, shape and texture, which rivals a chimpanzee's. People have sometimes speculated that the dolphin's large brain is necessary to house the intricate mental processes necessary for interpreting its surroundings by sonar. But a bat with a brain no larger than that of a mouse has a miniaturised sonar processor which works at least as well as a dolphin's.

Even within the evolutionary history of our own species the correlation between brain size and intelligence does not hold good. Most of us are happy to believe that as man evolved and his skull expanded he grew ever more and more intelligent. But most of us would be reluctant to believe that Neanderthal man was more intelligent than modern man simply because there were more cubic centimetres inside his skull. We might also find difficulty in accepting that the increase in human intelligence stopped dead when the increase in human skull size came to a halt.

So the Red Riding Hood approach is now less in evidence. It is the neoteny model which remains most influential — the concept that we are juvenilised versions of a more ape-like ancestor. The tempo of our growth rates has in some respects slowed down; we spend longer in the various stages of development between birth and death, and that includes spending longer in the infantile stage during which the brain is expanding very rapidly. The extended period of rapid brain growth means that we end up with relatively large skulls — a head/body ratio more typical of juvenile anthropoids than adult ones. (The neoteny factor and the ecological one are not, of course, mutually exclusive. Neoteny is a mechanism of evolutionary change; environmental conditions, such as changes in the food supply, could either facilitate the working of the mechanism or inhibit it.)

In 1989 C. P. Groves published a book entitled *A Theory of Human and Primate Evolution*, a detailed and meticulously documented account ranging from the earliest primate fossils to modern man. Near the end of it he declares himself ready to '. . . defend the following proposition: brain enlargement is an epiphenomenon of neoteny.'

This is the reverse of the more usual neoteny scenario, which suggests that the large brain somehow became so desirable and adaptive that other (disadvantageous) features such as hairlessness had to be tolerated as part of the package. Groves is suggesting that brain growth was '. . . not selected for as such', but happened fortuitously as a result of the changing tempo of our patterns of growth.

He then asks two very good questions. Most accounts of neoteny are content to compare modern *H. sapiens* with modern or primitive hominids. Groves raises the question of exactly when the neotenous trend began. He is inclined to regard it rather as Martin regards brain expansion — as a progressive phenomenon. A few neotenous features — such as the shortened face — appear very early; they are detectable, for example, in Lucy.

The second question he asks is why it happened. The probability is that a species does not become neotenous for no reason. The star example is that famous amphibian, the axolotl. It turns into a salamander when climatic conditions around its pond are sufficiently damp and shady to keep its skin moist. But if the surrounding land becomes too dry it retains its neotenous form, remains a tadpole all its life, and is capable of reproducing without becoming in other respects 'mature'.

Why, then, apparently at some time during the fossil gap, did man's ancestors take the first steps along the neotenous path? Groves does not attempt to answer the question. He merely points out that it exists, and he indicates the extreme complexity of the issues involved.

But in the course of his discussion he makes some interesting points, such as: 'The most advanced example of neoteny among mammals appears to be the order *Cetacea*' (that is, whales and dolphins). He cites a long list of resemblances between the general body form of an adult whale or dolphin and the form of the embryo of a land mammal such as a cow, pig or deer at the stage when the limb buds are forming. Like such an embryo, the dolphin has a hairless skin, a torpedo-shaped body with no neck, no external appendages such as ears, poorly formed ribs with no breast bone, hardly any hind-limb skeleton, and compared to most adult land mammals a very large brain in relation to body size — a characteristic of most mammal foetuses.

Groves speculated that the dolphin's large brain was neither adaptive nor maladaptive but 'exaptive' — a side-effect of the prolongation of foetal growth rates; it happened because in any mammal foetus at that stage the organ that is growing fastest happens to be the brain. He adds parenthetically: 'Doubtless all marine mammals are neotenous to some degree.'

It is not known why in mammals a move to an aquatic environment seems to be one of the triggers which sets in train a trend to neoteny. It is tempting to envisage it as an instance of *reculer pour mieux sauter*. Conceivably, a species finding itself in a radically new environment (such as water) begins to shed the more advanced features which fitted it for its old environment. It back-tracks to a more unspecialised foetus-like form, before re-adapting to the new habitat. If that were the case, then our own ancestors, having moved from the land to the water and subsequently from water to land, would have been subjected to an impetus towards neoteny on two successive occasions. It would explain why in our case the trend was unusually powerful.

All the hypotheses in this field are highly speculative; the problem is intensely complex and remains unsolved. The two possible connections between AAT and brain growth are those already indicated: (a) that a coastal environment, unlike a savannah one, would have provided the food resources without which brain growth could not have been afforded, and (b) that brain growth appears to be related to neoteny, and a trend towards neoteny is common among aquatic mammals but not among savannah ones.

Derek Ellis of the University of Victoria, Canada, has therefore analysed the ecology of all the various types of marine ecosystems found in tropical latitudes in relation to the observed behaviour of extant primates, with the object of assessing the viability of an aquatic ape. He examined the resources available in such environments as mangrove forests, salt-marsh lagoons, archipelagos, estuaries, reefs and rock shores, and collected data about swimming and other water activities in 26 primate species.

His starting point was a succinct encapsulation of the minimal AAT premise: 'We need a habitat other than a forest-savannah boundary ecotone for the separation of ape and human stocks.' His conclusion was that '. . . the tropical coastal environment provides productive ecosystems exploitable by a range of extant monkeys. There is no evident reason to believe that an ape could not converge to successful adaptation there.'

Yet this is precisely what many people find it hard to accept. Most of the attempts to refute AAT consist of alleged reasons why an aquatic ape could never have survived. It would, they imply, have been doomed to perish. It was not stream-lined; it was quite the wrong shape for a swimmer. It was warm blooded; once its fur was wet, the water would have drained it of its body heat. Its movements would be slow and inefficient; it could never move fast enough under water to catch a fish, and it would have starved. In the meantime its helpless young would have been greedily devoured by crocodiles and sharks, and it would have left no descendants.

These are all reasons why it would have been impossible for *any* land mammal to survive in water. They apply with equal force to the dog-like animal which went into the water and in the course of millions of years turned into a seal. Yet that primitive dog survived. Probably only some kind of duress would have driven it to resort to such a desperate measure, and doubtless many of the pioneers did perish. But enough of them survived, and today their descendants are found in oceans all over the world. It happened again and again — to the bear-like creature that turned into a walrus, and the mole-like creature that turned into a platypus, and the elephant's cousin that turned into a manatee, and the flying sea bird that turned into a penguin, and the ancient unknown quadrupeds that turned into whales and dolphins.

If the sea of Afar had not been cut off from the Indian Ocean we might have been able to add a primate to the list; some of the aquatic apes might have ventured into the open sea and stayed there while our own ancestors returned to the African continent.

It would have been fascinating to find cousins in the sea as genetically close to us as our cousins in the forest. But when the evaporating inland sea became untenably briny, any aquatic apes would perforce have gravitated to the inland shore where the water was fresher, and southward along the waterways, until ultimately the dependence on water as a food source was weakened and became dispensable.

Alister Hardy lived and died in the hope that some kind of proof might be dug up by the fossil hunters which would settle the matter. But it is hard to conceive of any fossil evidence that would be regarded as conclusive one way or the other. To take a parallel situation, if fossil hunters in five million years' time unearthed the fossilised skeletons of an otter and a stoat, they would be able to deduce that two closely related mustelids co-existed in Scotland in the twentieth century. They would find it very difficult to establish whether one of them was aquatic.

For a long time it appeared impossible to prove or disprove either the savannah theory or AAT. It seemed that belief in one or the other must always hinge on the balance of probability rather than anything more definite.

And when, finally, a new and quite unambiguous piece of evidence came to light it was widely ignored.

When the molecular biologists entered the field of evolutionary research, their first claim was that they could supply evidence about the timing of certain evolutionary events — the dates at which one species split off from another. During the 1970s their techniques were refined and developed, and they began to make statements about place as well as time. As G. J. Todaro declared: 'There is in fact a record of our history, however tenuous, that is recorded in all of us in our genes, and we are beginning to read that history.'

The best known example of a genetic marker giving evidence about our geographical origins is that of the haemoglobin S gene, which is of medical importance because it can sometimes cause sickle cell anaemia. This property is obviously maladaptive, and the normal expectation would be that it would have been bred out by natural selection. But it was discovered that it had another property which meant that in parts of Africa it *increased* the chances of survival: it provides a degree of resistance to malarial infection. Since the gene is not found in all human beings, it has obviously been acquired since the races of mankind dispersed and became differentiated. It is now fully accepted that the presence of the sickle cell gene is a marker which proves that its possessor must have had at least one African ancestor, though the family may have lived outside Africa for hundreds of years.

Far less familiar is the most challenging of all these markers. It is known as the 'baboon marker'. Its existence was first revealed in 1976 by a team of researchers from the National Cancer Institute in Bethesda, Maryland. Their paper was entitled: 'Baboons and their close relatives are unusual among primates in their ability to release nondefective endogenous type C viruses', and the authors were G. J. Todaro, C. J. Sherr and R. E. Benveniste. A subsequent article in *Nature* by Benveniste and Todaro had a more arresting title: 'Evolution of type C viral genes; evidence for an Asian origin of man'.

The infectious type C virus identified in the first paper belongs, like the AIDS virus, to the class known as retroviruses. When such a virus infects an animal, the RNA of the virus is converted into DNA inside the cells. This means that it becomes part of the genetic make-up of the infected animal and can be genetically transmitted from parent to offspring. Retroviruses can also be passed on and re-integrated into the DNA of other animals of the same species (as with AIDS), and even on rare occasions into the genetic information of distantly related species.

The type C virus is endogenous only in baboons — that is, it is a normal part of their make-up and has no harmful effect on them. But when it crosses the species barrier it has the potential to cause disease in other primates. The Bethesda team was able to establish that the baboon virus must at one time have constituted a threat to all the primates of Africa except the baboons themselves, because every African monkey and ape species they examined contained, in their chromosomes, viral gene sequences closely related to RNA genomes of the baboon virus; these sequences evolved to act as a kind of antibody, a protection against the baboon virus.

One thing we can deduce is that when the baboon virus first manifested itself it was able to cause — in all primates except themselves — a serious and probably life-threatening disease. Every extant African monkey or ape species carries the protective gene sequence in its chromosomes. If any such species failed to develop this resistance, it failed to survive.

The second thing we can deduce is that the virus was highly infectious. Unlike the AIDS virus, it was certainly not spread mainly by sexual contact. It was almost certainly airborne, because baboons are diurnal ground-dwellers, yet the species affected by the virus include not only other ground-dwellers like the chimpanzee and the gorilla, but also arboreal species like the colobus monkey and even small prosimians living high in the forest, like the bush baby and the galago.

Baboons still carry the virus, but it appears to have lost its virulence in the course of time. By now it constitutes no particular threat to other primate species, even those not carrying the protective marker. These are, of course, the South American and Asian primates whose ancestors were never in the vicinity of the baboons when the plague was at its height.

Among primates, then, the gene sequence (the 'baboon marker') is a reliable indication of African origin, just as the sickle-cell gene is in humans. Of the ape and monkey species tested, 23 showed the presence of the 'baboon marker' — and they were all African. They included the gorilla and the chimpanzee. Seventeen species, including the gibbon and the orang-utan, lacked the 'baboon marker' — and none of them were African.

The most remarkable aspect of Todaro's discovery emerged when he examined *Homo sapiens* for the 'baboon marker'. It was not there. All races of *Homo sapiens* — including the African races — lack the baboon marker.

Todaro drew one firm conclusion: 'The ancestors of man did not develop in a geographical area where they would have been in contact with the baboon. I would argue that the data we are presenting imply a non-African origin of man millions of years ago.'

His own choice of a non-African site was Asia. His paper implies that, following the split from the apes, some populations of hominids spread into Asia; that they were therefore absent when the baboon plague erupted; that any hominid population remaining in Africa at that time ultimately died out; and that all extant humans are descended from the Asian *émigrés*, some of whom returned westward and crossed the Suez isthmus back into Africa.

There is no evidence which actually rules out the hypothesis of an Asian origin. (The theory of descent from an 'African Eve' is irrelevant in this context; it refers to a population bottle-neck on the continent of Africa millions of years later, around 200,000 years ago.) Todaro's scenario, involving dispersal to the East followed later by a trek back to the West, was felt by some to be inherently improbable, but no one suggested any other way of accounting for the absence of the 'baboon marker' in man.

A non-African site does not necessarily imply Java or Peking. If the baboon plague broke out during the fossil gap when Afar was flooded, man's ancestors could have been living on Danakil — then an off-shore island miles away – or somewhere on the opposite shore of the proto-Red Sea. In either case, not even an airborne virus could have reached them, and by the time their descendants found themselves on the mainland, the baboon plague would have lost its virulence and they would never have needed to acquire the protective gene sequence.

The Bethesda findings made one thing perfectly clear. The idea that man's ancestors left the trees and subsequently spent all their time on the baboon-haunted African savannah has to be discarded. As Sarich said about molecular dating: 'One no longer has the option' of clinging to that version of events. Todaro knew that his findings would not be welcomed, but he wrote: 'As I see it, one of our major functions in the area of research is to act as historians — to report what actually happened rather than what we would have liked to happen.'

He was certainly not reporting what savannah theorists

like to think happened. Their reaction to the 'baboon marker' discovery was more or less the same as to Alister Hardy — the response was nil. Discussion of human evolution continued along the old familiar lines as if nothing had happened.

But there is a fundamental difference between trying to ignore a new hypothesis and trying to close the mind to new data. 'I don't wish to know that' was a favourite line of patter among cross-talk comedians, but no scientist can resort to it and hope to retain credibility.

In recent years the climate of opinion has begun to change. In 1987 an international conference was held at Valkenburg in the Netherlands, where the pros and cons of AAT were publicly debated, and that event represented some kind of watershed. By now a number of scientists are prepared to endorse the theory as a tenable hypothesis. Others, while remaining neutral, have been willing to offer constructive criticism, provide a platform, or suggest additional lines of enquiry. To all of these I owe a permanent debt of gratitude.

As Michael Ruse observed in a different connection: 'You can always find some way of propping up a crumbling theory but, if you are prepared to stay with science, then there really does come a time when the facts start to bite.'

For the crumbling savannah theory, that time has arrived.

References

ACKNOWLEDGEMENTS
Marc Verhaegen, 1987. The Aquatic Ape Theory and some common diseases, in *Medical Hypotheses*, **24**, 293–300.

CHAPTER 1
Elliot Smith, 1924. *Essays on the Evolution of Man*. Oxford University Press, p. 78.
Robert Broom, 1953. *The Coming of Man: Was it Accident or Design?* H. F. & B. Witherby, p. 10.

CHAPTER 2
Roger Lewin, 1987. *Bones of Contention. Controversies in the Search for Human Origins*. New York: Simon & Schuster.
Richard E. Leakey, 1981. *The Making of Mankind*. Michael Joseph, pp. 11–18.
Don Johanson and Maitland Edey, 1981. *Lucy: The Beginnings of Mankind*. Granada, p. 243.
Victor Sarich and Allan Wilson, 1967. Immunological time scale for hominid evolution, in *Science*, **158**, 1200–1203.
Sherwood Washburn, 1984. Quoted in Roger Lewin, *Bones of Contention, op. cit.*

David Pilbeam, Feb. 1984. The descent of hominoids and hominids, in *Scientific American*, p.87.

CHAPTER 3

Charles Darwin, 1871. *The Descent of Man*, chap. 2.

G. Schmorl, 1981. Quoted in Peter Medawar, 1981. *The Uniqueness of the Individual*. New York: Dover Publications.

W. M. Krogman, 1951. The Scars of Human Evolution, in *Scientific American*, **185**, (6), 54–57.

Roger Lewin, 1983. *Science*, **220**, 700–702.

CHAPTER 4

David Attenborough, 1979. *Life on Earth*. Collins, p. 294.

Jack Prost, 1985. Chimpanzee behaviour models of hominisation, in *Primate Morphophysiology, Locomotor Analysis and Human Bipedalism*, edited by Shiro Kondo, University of Tokyo Press, p. 299.

Michael H. Day, 1986. Origins and Modes, in *Major Topics in Primate and Human Evolution*, edited by Wood, Martin and Andrews, Cambridge University Press, chap 9, p. 199.

C. Owen Lovejoy, 1981. The Origin of Man, in *Science*, **211**, 341–350.

Graham Richards, 1986. Freed hands or enslaved feet? in *J. Hum. Evol.*, **15**, 148.

Russell Tuttle and David Watts, 1985. The positional behaviour and adaptive complexes of Pan gorilla, in *Primate Morphophysiology, op. cit.*

R. W. Newman, 1970. Why man is such a thirsty and sweaty naked animal: a speculative review, in *Human Biology*, **42**, 12–27.

Peter Wheeler, 1985. The loss of functional body hair in man: the influence of thermal environment, body form and bipedality, in *J. Hum. Evol*, **42**, 12–27.

W. Lawrence Bragg on Wegener. Quoted in *The Listener*, 19 July, 1984.

Don Johanson, 1981, *op. cit.* Chapter 2.

Paul Mohr, 1978, Afar, in *Ann. Rev. Earth Planet Sci.*, **6** 145–172.

Leon P. LaLumiere, 1981. The evolution of human bipedalism: Where it happened – a new hypothesis, in *Phil. Trans. R. Soc. London*, **B292**, 103–107.

Ernst Mayr, 1963. *Population, Species and Evolution.* Harvard University Press.

J. Hürzeler, 1960. The significance of *Oreopithecus* in the genealogy of man, in *Triang*, **4**, 164–174.

Oreopithecus is further discussed in papers, 1986, by Eric Delson; A. Azzaroli, M. Baccaletti, E. Delson, G. Moratti and D. Torr; Terry Harrison; Frederick S. Szalay and John Langdon, in *J. Hum. Evol.*, **15**, 164–174.

CHAPTER 5

Desmond Morris, 1967. *The Naked Ape.* Macdonald & Co.

William Montagna, 1982. The evolution of human skin, in *Advanced Views in Primate Biology*, Springer-Verlag, 35–42.

Ronald Marks, 1988. *The Sun and your Skin*, Macdonald & Co.

CHAPTER 6

Harry Nelson and Robert Jurmain, 1988. *Introduction to Physical Anthropology.* West Publishing Co, 4th edition.

Clifford J. Jolly and Fred Plog, 1987. *Physical Anthropology and Archaeology.* Alfred A. Knop, 4th edition.

V. E. Sokolov, 1982. *Mammal Skin.* University of California Press.

Louis Bolk, 1926. On the problem of anthropogenesis, in *Proc. Section Sciences Kon. Akad*, Wetens, Amsterdam, **27**, 465–475.

Stephen Jay Gould, 1977. *Ontogeny and Phylogeny.* Harvard University Press.

Gary G. Schwartz and Leonard A. Rosenblum, 1981.

Allometry of hair density and the evolution of human hairlessness, in *Am. J. Phys. Anthrop.*, **55**, 9–12.

R. M. Martin, 1977. *Mammals of the Seas*. B. T. Batsford.

W. Sokolov, 1962. Adaptations of the mammalian skin to the aquatic mode of life, in *Nature*, **195**, 464–466.

CHAPTER 7

J. S. Weiner and K. Hellmann, 1960. The sweat glands, in *Biol. Rev.*, **35**, 141–186.

J. S. Strauss and F. J. Ebling, 1970. Control and function of skin glands in mammals, in *Memoirs of the Society for Endocrinology*, **18**, 341–371.

Knut Schmidt-Nielsen *et al.*, 1957. Body temperature of the camel and its relation to water economy, in *Am. J. Phys.*, **188**, 188–189.

Derek Denton, 1982. *The Hunger for Salt. An Anthropological, Physiological and Medical Analysis*. Springer-Verlag.

W. Hancock and J. B. S. Haldane, 1939. The loss of water and salts through the skin and the corresponding physiological adjustments, in *Proc. Roy. Soc.*, **B105**, 43–60.

W. Montagna, 1972. The skin of non-human primates, in *Am. Zool.*, **12**, 109–124.

W. Montagna, 1982. The evolution of human skin, *op. cit.* Chapter 5.

William R. Keating *et al.*, 1986. Increased platelet and red cell counts, blood viscosity and plasma cholesterol levels during heat stress and cerebral thrombosis, in *Am. J. Med.*, **81**, 795–800.

CHAPTER 8

D. B. Dill *et al.*, 1933. Salt economy in extreme dry heat, in *Journal of Biological Chemistry*, **100**, 755–768.

C. L. Evans and D. F. G. Smith, 1956. Sweating responses in the horse, in *Proc. Roy. Soc.*, **B145**, 61–81.

Sheila A. Mahoney, 1980. Cost of locomotion and heat

balance during rest and running from 0° to 55°C in a patas monkey, in *Journal of Applied Physiology*, **49**, 61–81.

John K. Ling, 1965. Functional significance of sweat glands and sebaceous glands in seals, in *Nature*, **208**, 560–562.

R. Fange, K. Schmidt-Nielsen and H. Osaki, 1958. The salt gland of a herring gull, in *Biological Bulletin*, **115**, 162.

P. Shiefferdecker, 1917. Die Hautdrusen des Menschen, in *Biol. Zbl.*, **37**, 536–563.

William H. Frey, 1985. *The Mystery of Tears*. Harper & Row.

CHAPTER 9

Frederick Wood Jones, 1929. *Man's Place among the Mammals*. Edward Arnold.

Alister Hardy, 1960. Was man more aquatic in the past? in *New Scientist*, 642–645.

Caroline Pond, 1978. Morphological aspects and the ecological and mechanical consequences of fat depositions in wild vertebrates, in *Ann. Rev. Ecol. Syst.*, **9**, 519–70.

P. B. Medawar, 1955. The imperfections of man. Reprinted in *The Uniqueness of the Individual*, 1981, Dover Publications.

Caroline Pond, 1987. Fat and figures, in *New Scientist*, 4 June, 62–66.

P. F. Scholander *et al.*, 1950. Body insulation of some Arctic and tropical mammals and birds, in *Biol. Bull.*, **99**, 225–236.

CHAPTER 10

A. Cryer and R. L. R. Van, 1985. *New Perspectives in Adipose Tissue: Structure, Function and Development*. Butterworth.

A. Häger, 1981. Adipose tissue cellularity in childhood in relation to the development of obesity, in *Brit. Med. Bull.*, **37**, (3), 287–290.

B. Larsson *et al.*, May 1984. Abdominal adipose tissue

distribution, obesity, and risk of cardiovascular disease and death: 13 year follow up of participants in the study of men born in 1913, in *BMJ*, 1401–1404.

John Studd *et al.*, March 1985. *Brit. J. of Obstetrics and Gynaecology*.

CHAPTER 11

Victor E. Negus, 1929. *The Mechanism of the Larynx*. Heinemann.

Victor E. Negus, 1965. *The Biology of Respiration*. E. & S. Livingstone.

Christian Guilleminault *et al.*, 1980. Sleep apnea syndrome: recent advances, in *Internal Medicine*, **26**, 347–372.

Edward Crelin, Can the cause of SIDs be this simple? in *Patient Care*, **12**, 5.

G. A. de Jonge and A. C. Engelberts, 1989. Cot Deaths and Sleeping Position, in *The Lancet*, 11 Nov., **8672**, 1149–1150.

Michael Campbell and Mike Murphy, March 1987. Cot deaths following thaw, in *Journal of Epidemiology*.

F. Wood Jones, 1940. The nature of the soft palate, in *J. Anat.*, **74**, 147–70.

Jan Wind, 1976. *Phylogeny of the Human Vocal Tract*. New York Academy of Sciences.

M. R. Mukhtar and J. M. Patrick, 1984. Bronchoconstriction: a component of the 'diving response' in man, in *Eur. J. Applied Physiology*, **53**, 155–158.

M. R. Mukhtar and J. M. Patrick, 1986. Ventilatory drive during face immersion in man, in *J. Physiol.*, **370**, 13–24.

R. van den Berg and J. Wind, 1987. Has man's upright posture contributed to speech origin by lowering the larnyx? in *Clin. Otolaryngol.*, **12**.

Robert Elsner and Brett Gooden, 1983. *Diving and Asphyxia*. Cambridge University Press.

M. J. H. van Bon *et al.*, 1989. Otitis media with effusion

and habitual mouth breathing in Dutch pre-school children, in *Int. J. Pediatri. Otorhinolaryngol.*, **17**, 119–125.

CHAPTER 12

R. D. Martin and R. May, 1981. Outward signs of breeding, in *Nature*, **293**, 7–9.

A. Harcourt *et al.*, 1981. Testis weight, body weight, and breeding systems in primates, in *Nature*, **293**, 55–57.

C. S. Ford and F. A. Beach, 1952. *Patterns of Sexual Behaviour*. Eyre & Spottiswoode.

K. E. Fichtelius, 1988. Contribution to the discussion on bipedalism, in *OSSA*, **14**, 45–49.

Ronald D. Nadler, 1981. Laboratory research on sexual behaviour of the great apes, in *Reproductive Biology of the Great Apes: Comparative and Biomedical Perspectives*, edited by Charles E. Graham. Academic Press.

B. M. F. Galdikas, 1981. Orang-utan reproduction in the wild, in *Reproductive Biology of the Great Apes, op. cit.*

CHAPTER 13

Stephen Jay Gould, 1977. *Ever Since Darwin*. Penguin Books, 160–167.

Max Westenhöfer, 1942. *Der Eigenweg des Menschen*. Mannstaedt & Co.

Michael Ruse, 1989. *The Darwinian Paradigm*. Routledge.

Kerstin Schuitema, 1990. The significance of the human diving reflex. Paper presented at a Symposium on Human Evolution — *The Aquatic Ape: Fact or Fiction* — Valkenburg, Netherlands, 28–30 August, 1987.

Donald W. Rennie, quoted in The Diving Women of Korea and Japan, by S. K. Hong and Hermann Rahn, 1967, *Scientific American*, **216**, (5).

CHAPTER 14

R. D. Martin, 1983. Human brain evolution in an ecological context, in *Am. Mus. Nat. Hist.*

Michael Crawford and David Marsh, 1989. *The Driving Force*. Heinemann.

C. P. Groves, 1989. *A Theory of Human and Primate Evolution*. Oxford University Press.

Derek Ellis, 1990. Is an aquatic ape viable in terms of marine ecology and primate behaviour? Paper presented at a Symposium on Human Evolution, *op. cit*, Chapter 13.

G. J. Todaro, 1980. Evidence using viral gene sequences suggesting an Asian origin of man, in *Current Arguments on Early Man*, Pergamon Press.

Index